HEIDELBERG SCIENCE LIBRARY | Volume 13

John C. Eccles

Facing Reality

Philosophical Adventures by a Brain Scientist

With 36 Figures

Springer-Verlag New York · Heidelberg · Berlin 1975

Professor Sir John C. Eccles,
Department of Physiology,
State University of New York at Buffalo
School of Medicine and Dentistry, Buffalo, NY 14214/USA

This is a reprint of the first edition published 1970.

ISBN 0-387-90014-4
ISBN 3-540-90014-4

For Helena

Preface

The titling of this book — "Facing Reality" — came to me unbidden, presumably from my subconscious! But, when it came, it seemed to be right, because that essentially is what I am trying to do in this book. "Facing" is to be understood in the sense of "looking at in a steadfast and unflinching manner". It thus contrasts with "Confronting" which has the sense of "looking at with hostility and defiance". As I face life with its joys and its sorrows, its successes and its failures, its peace and its turmoil, my attitude is one of serene acceptance and gratitude and not one of angry and arrogant confrontation and rejection.

The other component of the title — "Reality" — is the ultimate reality for each of us as conscious beings — our birth — our self-hood in its long stream of becoming throughout our life — our death and apparent annihilation. This is the Reality that we each of us must face if we are to live and adventure as free and responsible beings and not as mere playthings of chance and circumstance, going through a meaningless farce from birth to death with the search ever for distraction and self-forgetfulness.

As a brain scientist I have specialist knowledge of that wonderful part of the body that is alone concerned in the whole life-long interplay between the conscious self and the external world, including other selves. There will be a critical account of the way the brain receives from the world and the way it acts on it, and also of the possible synaptic mechanisms of memory — both storage and retrieval. But more importantly there will be a philosophical treatment of the age-old problems of the liaison between the brain and the conscious self, as uniquely known directly to each one of us, with its amazing continuity and unity. And this relationship of brain to conscious self in turn leads on to a consideration of the religious concept of the soul and the possibility of a deeper meaning in personal life that we at present may not be able to discern after the last century of philosophical and religious catharsis.

My life-long experience as a scientist has prompted me to discuss the nature of science and the making of scientists. It can be claimed that in our civilization science represents the highest creative activity of man, and it provides our best hope of winning through to a deeper understanding of the nature and meaning of existence. Philosophy must again become receptive to the ongoing scientific disputations; hence scientists have to dare to talk philosophically, though we must be attentive to the criticisms of those philosophers who are searching also for a common language to express a philosophy that is fully cognizant of science.

After the radiance and optimism of the first few years of this century, European civilization has been overwhelmed by a succession of disasters with only brief respites, as in the twenties. I look with forboding at the future of our European civilization. Violent irrationalism captures and enslaves so many of the so-called intellectuals — old as well as young. I reject all this darkness and evil that seems to overwhelm us. However, in this book, my whole effort has been directed positively in the attempt to build upon the magnificent heritage of our civilization with its rationality and its beauty. My hope is that I may restore to my fellow men a sense of the wonder and mystery of their own personal existence on this beautiful planet that is ours. I hope that I can inspire them with the courage to adventure wisely in achieving a new illumination in the last decades of this turbulent century.

My very good friend, Doctor HEINZ GÖTZE, President of Springer-Verlag, encouraged me to collect various lectures and edit them for publication in the Heidelberg Science Library. In addition I have added various sections and even some whole chapters of unpublished material in order to give some form to the ensemble. I wish to acknowledge with gratitude the permissions kindly given to reprint from previously published lectures as noted in the beginning of some chapters.

My wife, HELENA, has helped me in every aspect of this work with the texts, with the figures and with the indexing. I wish to thank my assistant, Miss VIRGINIA MUNIAK, for the preparation of the manuscript, in which she was helped by Mrs. CLAUDIA LEY, and Mr. JOSEPH WALDRON for the photography.

I express my grateful thanks to Doctors BARONDES, COLONNIER, HÁMORI, HELD, HUBEL, JANSEN, JUNG, LIBET, LØMO, PALAY, PENFIELD, PHILLIPS, SPERRY, SZENTÁGOTHAI and VALVERDE for kindly allowing me to reproduce figures from their papers and for making original figures available for reproduction.

My special thanks go to the publishers for their unfailing courtesy and for their extraordinary efficiency.

Buffalo, July 1970 John C. Eccles

Table of Contents

Acknowledgments

Grateful thanks are due to the publishers and editors of the following journals for their generosity in giving permission for reproduction of figures: Acta Biologica Hungarica, Brain Research, The Clarendon Press, Experimental Brain Research, Experimental Cell Research, Journal of Comparative and Physiological Psychology, Journal of Physiology, Perspectives in Biology and Medicine, Proceedings of the National Academy of Sciences, Progress in Brain Research, Science, Scientific American.

Chapter I

Introduction: Man, Brain and Science[1]

All men of good will would subscribe to the concept that we must strive to foster and develop the fullest possible life for mankind, not just here and now, but indefinitely into the future, as has been so eloquently expressed by DUBOS (1968). It is my belief that we will be successful only insofar as we appreciate the nature of man and plan accordingly. Man is self-reflecting in that he has the ability to objectify himself and to consider the kind of being he is and what he wants to become. Man alone is conscious of himself and is alone capable, as it were, of standing outside of himself and regarding himself as an object. As I come to consider the nature of man, I discover that I have direct access to privileged information about one—namely myself with my self-consciousness. I am not going to use this assertion in order to develop a solipsistic thesis. I shall be at pains to show that I have to recognize an equivalent self-consciousness in all other human beings. My philosophical position (cf. ECCLES, 1965a, 1965b, 1969a) is diametrically opposite to those who would relegate conscious experience to the meaningless role of an epiphenomenon.

Is it not true that the most common of our experiences are accepted without any appreciation of their tremendous mystery? Are we not still like children in our outlook on our experiences of conscious life, accepting them and only rarely pausing to contemplate and appreciate the wonder of conscious experiences? For example, vision gives us from instant to instant a three-dimensional picture of an external world and builds into that picture such qualities as brightness and color, which exist only in perceptions developed as a consequence of brain action. Of course we now recognize physical counterparts of these perceptual experiences, such as the intensity of the radiating source and the wave lengths of its emitted radiation; nevertheless, the perceptions themselves

1 This section is modified from a publication in Perspectives in Biology and Medicine (ECCLES, 1968).

arise in some quite unknown manner out of the coded information conveyed from the retina to the brain. Perhaps it is easier still to appreciate the miraculous transformation that occurs in hearing—from mere congeries of pressure waves in the atmosphere to sound with tone and harmony and melody. These sensory experiences arise when fleeting patterns of neuronal activities develop in the brain in response to the inflow of nerve impulses providing coded information from the auditory mechanism in our ears. These patterns are woven in space and time by the transient activation of cerebral nerve cells. There are over ten thousand million of these cerebral nerve cells, and the virtually illimitable possibilities of connectivity between them give the potentiality for an almost infinite variety of patterned operation. There is good reason to believe that spatio-temporal patterns involving tens of millions of these cells must be activated before we experience even the simplest sensation (Chapters IV, V).

I hope these simple examples convey some impression of what I mean when I refer to the wonder of the conscious life that each of us experiences. Yet it seems to me that post-Darwinian man has in this age lost the sense of his true greatness and of his immeasurable superiority to animals. Mankind is sick and has lost faith in itself and in the meaning of existence. There are many symptoms of this sickness or alienation. It has resulted in various forms of irrationality, such as existentialism in philosophy and the meaninglessness and formlessness of so much of so-called modern art. This occurs not only in the plastic arts but also in music and literature. Many other symptoms are exhibited by the adolescents of our day: their revolts against all traditions, the meaninglessness of their behavior patterns, their irrationality, and their propensity to go on psychedelic "trips" to enlarge their consciousness. And the adults display the pervading hedonism of life and a sense of values that is based entirely upon the purposeless gaining and spending of wealth. I would suggest that this disease is very widespread and more serious than anything with which mankind has ever been afflicted in the past.

In order to appreciate the situation in which man now finds himself, we must be cognizant of his past from his earliest evolutionary origin right up to the present time. We have to study the myths and religions by which man has lived and developed. He was thus able to gain an assurance and deep respect for the meaning and significance of his life and a faith in his destiny. On this great canvas of history we can examine the immediate conditions that seem to be responsible for his present unhappy state.

Just over 100 years ago the evolutionary story of man's origin was developed, so that for most thinking men it soon became certain that man was not some being of special creation by God. He had evolved by a process which can be thought of as a dialogue between gradual genetic change and the rigorous operation of natural selection, whereby all unfavourable developments were ruthlessly weeded out by what was called "survival of the fittest". Though it is over 100 years since this theory of evolution was first proposed in its full implications for man, the impact of it on the emotional life of thoughtful men took many decades, involving a process of action and reaction. The evolutionists often claimed that evolution provided a complete explanation of man's origin and was established with scientific certainty. Conversely, those committed to a religious view of man went on record as denying completely the evolutionary story so far as man was concerned. Gradually over the last decades this dispute has filtered down to the masses, greatly disturbing their often naive religious beliefs. It seems that mankind in general is suffering more than ever from the psychological trauma of this violent dispute.

There has been over the last decades a distortion of psychology into a purely behavioural psychology of a strictly deterministic and so-called objective character. This psychology categorically stated that all deliverances of consciousness were subjective, and hence were meaningless in relation to the scientific project to understand the behaviour of man within the framework of a deterministic psychology. There was, of course, the implication that our sense of purpose and decision was an illusion and that we were caught up in a rigorous web of determinism that was inexorably governed by just two factors: our inheritance and our conditioning. From this purely deterministic psychology there developed an irresponsibility and a feeling of the meaninglessness of life. We suffer today from the unjustifiable violence done by psychologists to the science of psychology when they reject from its subject matter all conscious experiences.

At this present primitive level of understanding of man's brain and of its role in ordering behaviour, it is often maintained that man is merely a special type of computer. I will readily agree that this can occur in special regions of the brain. For example, in recent years my research has been concentrated on the cerebellum, which is a highly differentiated organ of the brain especially concerned with both the control and the finesse of movement. As I try to understand how the cerebellum performs this task, I am virtually constrained to think of its acting essentially as a computer, but one whose principles of operation radically differ from present-day computers. However, the cerebral cortex contrasts with the cerebellum by the immeasurably greater complexity of its structure and functional performance. Moreover, we have to recognize that certain highly complex patterned actions of the cerebral cortex and the associated subcortical ganglia give rise to conscious experience, whereas there are no grounds for believing that conscious experiences ever arise in the cerebellum with its relatively simple computer-like performance.

There is no evidence whatsoever for the statements often made that, at an adequate level of complexity, computers also would achieve self-consciousness. To that extravagant claim there may be added the further assertion, by these same people, that computers may achieve a higher status of evolutionary development than man, so that we would be relegated to some servile role, just as animals are by us. When it is asserted that a sufficiently complex computer will achieve self-consciousness, there is a failure to recognize that even extremely complex cerebral activities often do not result in a conscious experience. As yet we are quite ignorant of the special circumstances which attend those patterns of neuronal activities in the brain that are giving rise to conscious experiences (Chapters IV, V).

Again, we have the extraordinary doctrine that all behaviour is determined solely by inheritance and conditioning (cf. Chapter VIII). It is unfortunate that the advocates of this doctrine have been blind to the fact that its logical consequences make its assertion meaningless; for their act of assertion should be recognized by them as merely the result of a prior conditioning, thus signaling merely the effectiveness of this conditioning! The denial of free will and the advocacy of a universal determinism have been asserted within the scientific framework both of a primitive type of reflexology as an epitome of brain performance and of the now-discredited nineteenth-century deterministic physics. By logical analysis both POPPER (1950) and MACKAY (1966) have shown that even deterministic physics does not render untenable our belief in the freedom of the will. My position is that I have the indubitable experience that by thinking and willing I can control my actions if I so wish, although in normal waking life this prerogative is exercised but seldom. I am not able to give a scientific account of how thought can lead to action, but this failure serves to emphasize the fact that our present physics and physiology are too primitive for this most challenging task of resolving the antinomy between our experience and the present primitive level of our understanding of brain function. When thought leads to action, I am constrained, as a neuroscientist, to postulate that in some way, completely beyond my understanding, my thinking changes the operative patterns of neuronal activities in my brain. Thinking thus comes to control the discharges of impulses from the pyramidal cells of my motor cortex and so eventually the contractions of my muscles and the behavioural patterns stemming therefrom.

There is general agreement among neuroscientists that every conscious experience—every perception, thought, and memory—has as its material counterpart some specific spatio-temporal activity in the vast neuronal network of the cerebral cortex and subcortical nuclei, that is woven of neuronal activities in space and time in the "enchanted-loom" so poetically described by SHERRINGTON (1940). I would further go on to say that, no matter what one's philosophical or political position, there must be general assent to the proposition that the study of the brain is central to the scientific investigation of the nature of man. The functioning of the brain gives us all that matters in life, not only our immediate perceptions, as I have illustrated by vision and hearing, but also all memory, emotions, thoughts, ideals, imagination, technical skills, and above all the creative achievements in art, philosophy, and science (Chapters IV, V, and X).

In the last ten to twenty years there has been enormous progress in the understanding of the simpler aspects of brain structure and function (cf. Chapter II). This basic work has been on the properties of the unit structures of the nervous system, the nerve cells, on the modes of communication over nerve cells by propagated impulses and between them at the functional contacts or synapses, and on the simplest functional patterns of nerve cell organization. These discoveries provide a sound foundation for further progress; and, thanks to the great power of the new microtechniques, the successes have been beyond the most sanguine hopes of a few years ago. We can now be confident that an almost infinite complexity and variety can be created in the patterning of the neuronal network. This dynamic patterning of operation by the ten thousand million nerve cells of the cerebral cortex would endow it with potentialities adequate for any achievement, even for that displayed in the performance of the highest intelligence. I have hopes, too, that we are on the threshold of understanding the basic principles responsible for the laying down of memory traces, which we may envisage as being due to an enduring enhancement of synaptic efficacy with usage (cf. Chapter III). In this way a neuronal pathway that is activated by a particular sensory input will, as a consequence of repeated activation, achieve a kind of stabilization by means of the enhanced synaptic functions of its neuronal linkages. This patterned engram, as it is called, is available for recall in memory when there is an appropriate input into its circuitry.

Yet all this progress serves only to give an immensely wider and deeper vision of the fantastic problems that lie ahead. If we are to regard the brain as a machine, it enormously transcends in its variety and flexibility any man-made machine, such as a computer. If it is to be regarded as a communication system, such as an automatic telephone exchange, it is of a quite different order from anything that we can devise, comprising, as it does, more than ten thousand million neurones linked up in the most unimaginably complex and subtle manner. And, as I have already emphasized, it has in addition the extraordinary emergent property of providing self-consciousness, at least in certain states of activity.

There need be no fear that this attempt to understand the brain scientifically will lead to the removal of the "final illusions of man about his own spiritual existence," which would be the claim of some positivist scientists as well as philosophers. On the contrary, the framework of a quite inadequate and primitive concept of the brain provides the medium in which flourish the materialistic, mechanistic, behaviouristic, and cybernetic concepts of man, which at present dominate research. Of course, I fully support scientific investigations on behaviour and conditioned reflexes and, in fact, all the present scientific programs of behaviouristic psychology. Furthermore, I agree that much of human behaviour can be satisfactorily explained on the basis of concepts developed in relation to these experiments. However, I differ radically from the behaviourists in that they claim to give a *complete* account of the behaviour of a man, whereas I know that it is not adequate to provide me with an answer to the question: What am I? It does not explain me to myself, for it ignores or relegates to a meaningless role my conscious experiences, and to me these constitute the primary reality—as doubtless it does to each one of you, my readers (cf. Chapter X).

I have an equivalent objection to the assertion of evolutionists that my brain, as well as my consciously experiencing self is *fully* explained by the magnificent creative process of thousands of millions of years of evolution. I readily accept all that they postulate in respect of my brain, yet I find my own consciously experiencing self not satisfactorily accounted for. The creative evolutionary process is to me an incomplete explanation of the origin of myself (Chapter VI). I believe that we have to recognize that there are great unknowns in the attempts that have so far been made to understand the nature of man. And the further we progress in research, the more each of us will realize the tremendous mystery of our personal existence as a consciously experiencing being with imagination and a sense of values and a systemization of knowledge. Man's story and his whole knowledge can be stored and transmitted in the coded form of written language; and so knowledge develops progressively, giving modern man an amazing power to understand nature and hence to control it. Though an animal and the ultimate product of evolution, he transcends animals so that he can be regarded as a different order of being with his ideals, his art, his values, his science, and, above all, his self-consciousness (cf. Chapter X).

If this concept of man could be brought home to the artists and writers so that they could emotionally feel its tremendous impact upon life, I am confident that they would be inspired again by the uniqueness, wonder, beauty, and dignity of human life. For it seems to me that this can now be better appreciated than ever before.

In relation to brain research, it is important that the actual nature of scientific investigation be appreciated. Even eminent scientists (cf. CRICK, 1966; STENT, 1969) sometimes err in telling people that science is giving a fundamental understanding of nature that will soon be complete at least in essentials and that life and the whole of our conscious existence inevitably will be reduced to physics and chemistry. I have no objections to scientists expressing these views so long as they do not claim to speak with the authority of Science, which is assumed by the public to provide certainties that must be accepted unreservedly. We must instead recognize that science is in fact a personal performance of scientists, each explaining some aspect of nature and expressing these explanations to others for their critical judgement and experimental testing. In this way there can be a progressive elimination of error from the imaginative hypotheses, and so a progressive approximation to truth, although it must be realized that truth in itself can never be fully attained except at a trivial level (cf. POPPER, 1962). The most we can claim to do in our attempts to understand nature (Chapter VII) is to be able to develop hypotheses that approach progressively nearer to truth, and this limitation includes particularly that most difficult of all scientific tasks, the understanding of the brain.

In fact, science is shot through with values—ethics in our efforts to arrive at truth and aesthetics in our conceptual imagination and in the appreciation of our hypotheses. If we can give to mankind an understanding of science as a very human endeavour to understand nature and to present in all humility the best of our feeble efforts to do so, then science would be appreciated as a great and noble human achievement, whereas, instead, it is in danger of becoming some great monster feared and worshipped by man and carrying with it the threat to destroy man.

Chapter II

The Neuronal Machinery of the Brain

Before attempting to give an account of ideas and problems relating on the one hand to the neuronal mechanisms of the brain and on the other to conscious experience, it is essential to give a brief and very elementary account of some properties of the nervous system in order to provide a simple structural and functional basis for these accounts.

The human cerebral cortex is the great folded structure that forms the major part of our brain and is the highest level of the evolutionary development of the nervous system, as may be seen in Fig. 1, where the brains of

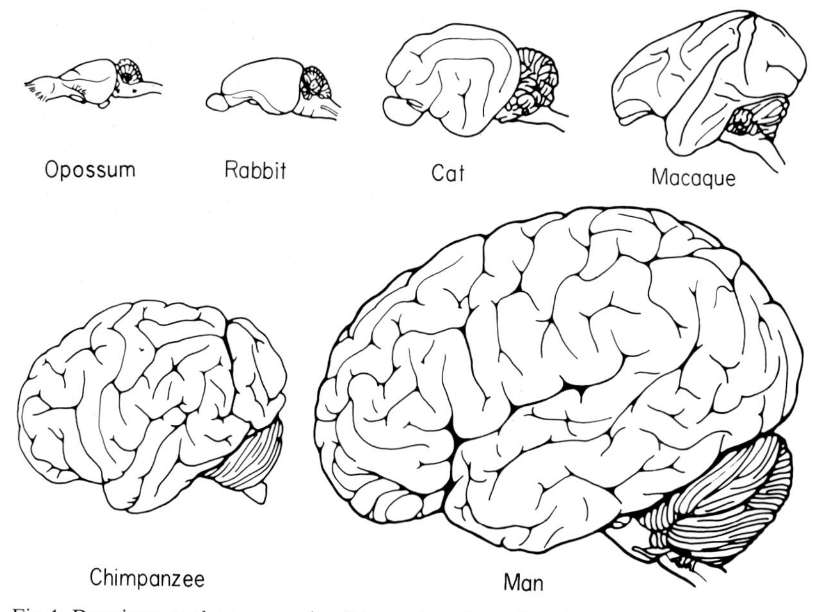

Fig. 1. Drawings on the same scale of the brains of a series of mammals. (Figure kindly provided by Professor J. JANSEN.)

some other mammals are shown on the same scale. Some idea of its size can be gained by thinking of a sheet of material 50 cm square and three millimeters thick, which approximates to the dimensions of our cerebral cortex, if it were unfolded and straightened out. As shown in Fig. 2, the

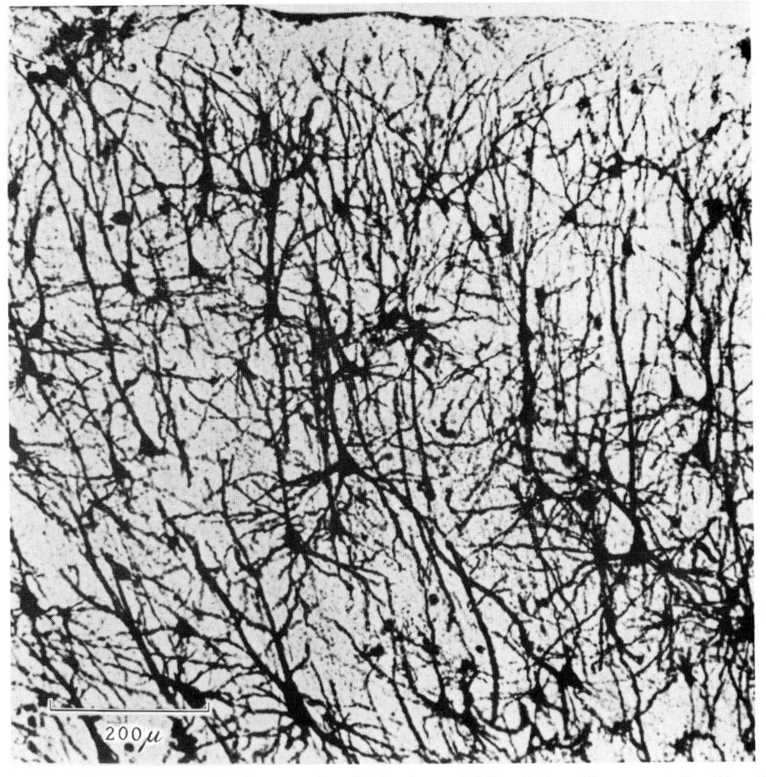

Fig. 2. Section of the cerebral cortex in which about 1.5% of the cells were stained by the Golgi-Cox method. Note the many large pyramidal cells with their branched dendrites (personal communication from the late D. A. SHOLL)

cerebral cortex is formed by a very dense packing of the unitary components of the nervous system which are called nerve cells, and which are of many different varieties and sizes, there being in all about 10,000-million such individual components in a human cerebral cortex (THOMPSON, 1899).

The most important generalizations that can be made about the central nervous system are that it is composed of an immense number of individual nerve cells or neurones and that these individuals are organized into functional assemblages by the synaptic contacts that they make with each other. For example, Fig. 3A shows a drawing by RAMÓN Y CAJAL

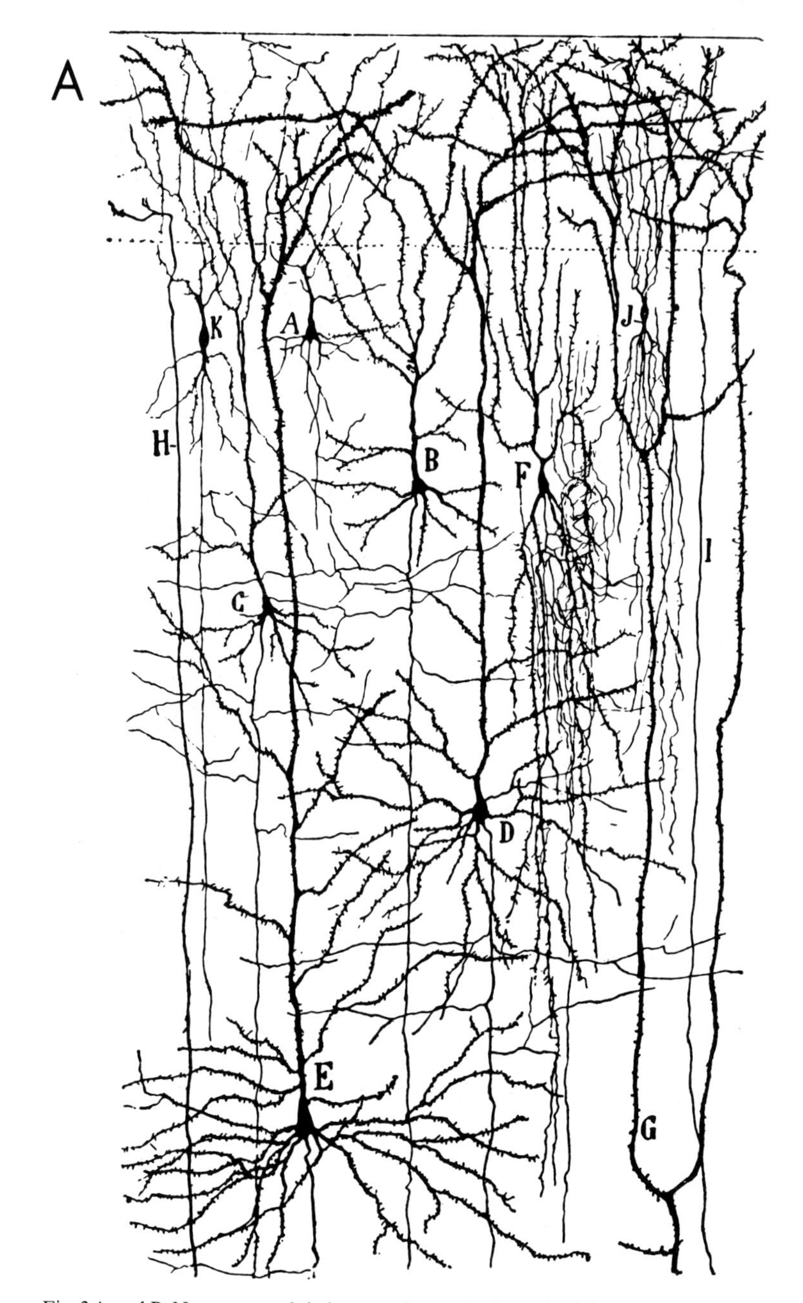

Fig. 3 A and B. Neurones and their synaptic connections. A Eight neurones are drawn from a Golgi preparation of the three superficial layers of frontal cortex from a month old child. Small (*A*, *B*, *C*) and medium (*D*, *E*) pyramidal cells are shown with their profuse dendrites covered with spines. Also shown are three other cells (*F*, *J*, *K*), which are in the general category of Golgi type II with their localized axonal distributions. *G* – *I* are dendrites from

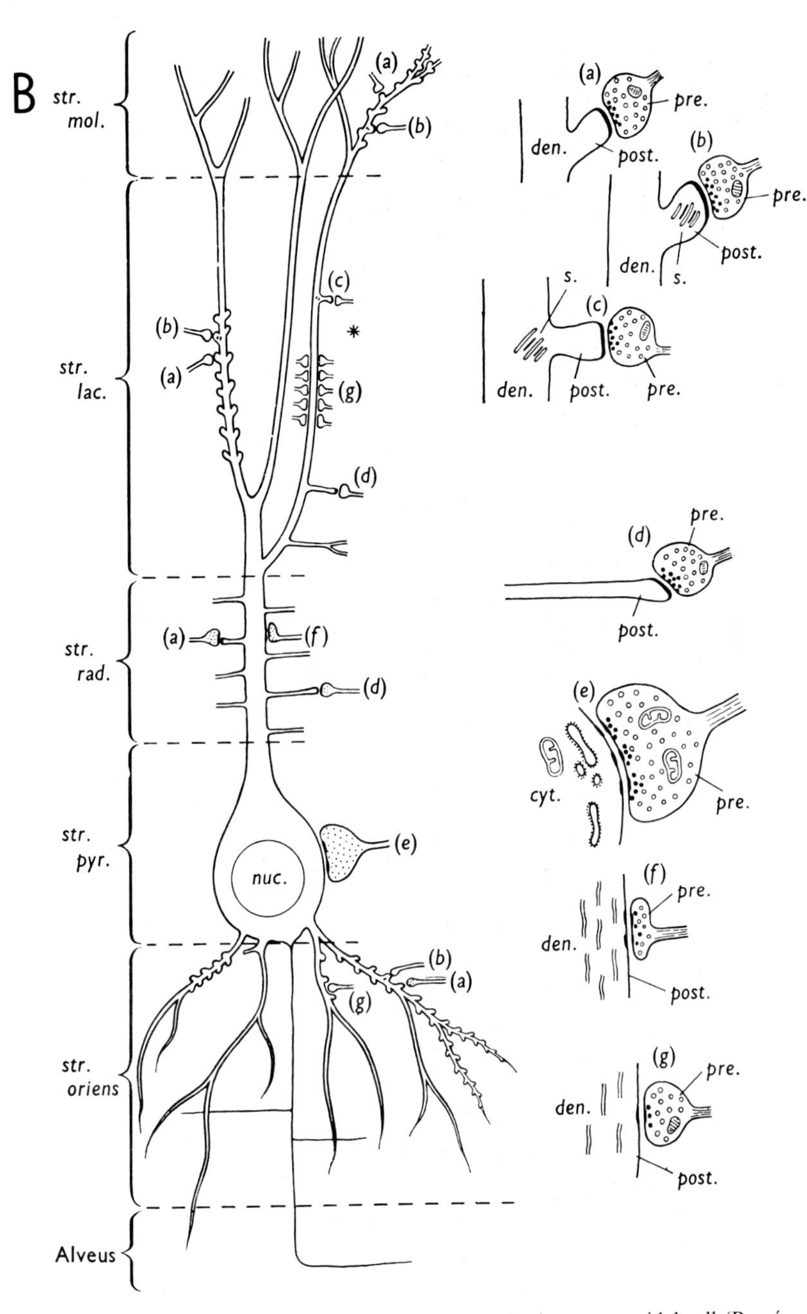

deeper lying neurones, G being the apical dendrite of a large pyramidal cell (RAMÓN y CAJAL, 1911). B Drawing of a hippocampal pyramidal cell to illustrate the diversity of synaptic endings on the different zones of the apical and basal dendrites, and the inhibitory synaptic endings on the soma. The various types of synapses marked by the letters *a*–*g* are shown in detail to the right (HAMLYN, 1963)

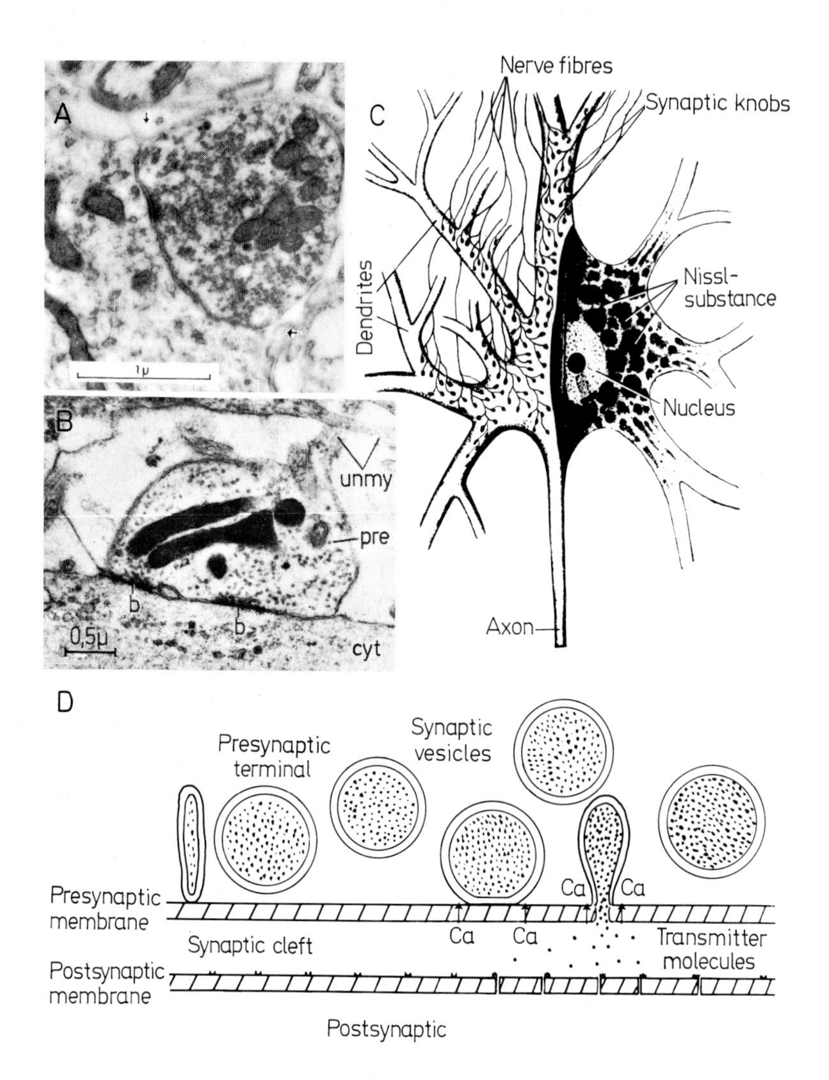

Fig. 4. A Electron-micrograph by Palay (1958) of a synaptic knob separated from subsynaptic membrane of a nerve cell by a synaptic cleft (marked by arrows) about 200 Å wide. In some areas the vesicles are seen to be concentrated close to the synaptic surface of the knob, and there is an associated increase in membrane density on either side of the cleft. B Electron-micrograph by Hamlyn (1963) of an inhibitory synapse that is formed by a synaptic knob (pre) of a basket cell on the soma (cyt) of a hippocampal pyramidal cell, there being two active sites (b). C Diagrammatic drawing by Jung of a neurone showing dendrites and axon radiating from the cell body or soma that contains the nucleus. Several fine nerve fibers are shown branching profusely and ending in synaptic knobs on the body and dendrites. D Schematic drawing of synaptic cleft

(1911) of eight neurones in the three most superficial layers of the frontal cortex of a month-old child. The body or soma of each neurone gives off elaborately branching dendrites covered with a wealth of small spines, and also a single fine axon process having many branches that are destined to make synaptic contacts on other neurones. No synaptic connections are seen in Fig. 3A, but a wide variety is shown in Fig. 3B that was drawn by HAMLYN (1963) on the basis of electron microscopic pictures. The small terminal branches from other neurones are shown making close contacts (synapses) with the soma and dendrites of this pyramidal cell of the hippocampal cortex, and the details of synaptic structure are shown in the various enlarged drawings to the right of Fig. 3B and in the electron-micrographs of Fig. 4A and B. For our present purpose it is sufficient to recognize that there is a complete separation between the presynaptic terminals and the postsynaptic membrane on which they make contact, there being actually a narrow separating space, the synaptic cleft, the line of which is indicated by arrows in Fig. 4A, and shown enlarged in Fig. 4D.

Furthermore Fig. 3B shows the elements of the chemical transmitting mechanism of the synapse. When the electrical message or nerve impulse propagates up to the presynaptic terminal, transmission across the synapse is not effected by an electrical mechanism, but by transmutation to a most ingenious chemical mechanism. The specific chemical transmitters are prepackaged in quanta of some thousands of molecules that are believed to be contained in the vesicles (cf. Fig. 4A, B, D) of the presynaptic terminal. The arriving impulse causes one or a very few vesicles to liberate their quanta of transmitter into the synaptic cleft (Fig. 4D). Thus the transmitter can act on special receptor sites of the postsynaptic membrane (Fig. 4D) opening specific gates for such ions as sodium, potassium and chloride that diffuse across the membrane and so generate a potential change across it. Two oppositely acting types of synapse can be distinguished—the excitatory in which ionic gates for sodium and potassium are opened, and the inhibitory in which the gates are for potassium and chloride (Fig. 26; cf. ECCLES, 1964, 1966d).

The operation of excitatory synapses is illustrated in Fig. 5. As shown in the inset, an electrode can stimulate many nerve fibers converging onto the motoneurone (cf. Fig. 4C) that is being recorded from by means of a microelectrode inserted into the soma so that there is accurate and selective observation of the electrical potential changes produced by synaptic action on that neurone. Only when the synaptic excitatory action is above a critical level does it produce a depolarization large enough to generate the action potential that causes the discharge of an impulse down its axon in the ventral root. For example in the upper trace of C there is only the small relatively slow depolarization (upward deflection) of the excitatory postsynaptic potential (EPSP), whereas in B the larger EPSP generates the large action potential resembling the action potential of A which is produced by firing an impulse from its axon (see electrode

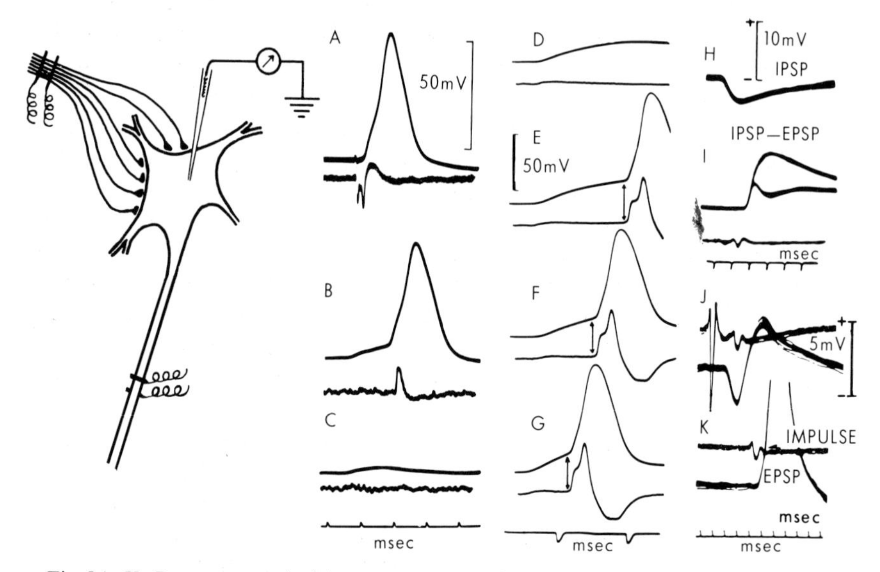

Fig. 5A–K. Responses evoked in a motoneurone by excitatory and inhibitory synaptic action. There is a schematic illustration of the motoneurone with intracellular electrode and electrodes on its monosynaptic excitatory input (from the nerve to its muscle) and its motor axon (in a ventral root filament). In A is the antidromic spike potential of that motoneurone generated by stimulation of its axon, while in B and C the stimulus is applied to the monosynaptic afferents, being above threshold for impulse discharge in B and just below in C. The lower traces in B and C are from the ventral root filament. (Membrane potential, −48 mV). D to G are intracellular recordings from another motoneurone in which monosynaptic excitation was progressively increased from D to G. The lower trace is the first differential of the upper trace, the double-headed arrows indicating the instant of impulse generation. Membrane potential, −70 mV (Coombs, Curtis and Eccles, 1957). H is IPSP set up in a biceps-semitendinosus motoneurone (membrane potential, −70 mV) by a quadriceps afferent volley and in I it is superimposed on the monosynaptic EPSP evoked in that motoneurone by an afferent volley from its own muscle, the EPSP being shown alone in half the superimposed traces (Curtis and Eccles, 1959). In J and K the lower traces are intracellular records similarly evoked in another motoneurone (membrane potential, −66 mV), but in K the EPSP evokes an impulse discharge, and in J this discharge is invariably inhibited by a preceding IPSP. Upper traces are the monitored records from the afferent volleys entering the dorsal root (Coombs, Eccles and Fatt, 1955)

in inset diagram) into the nerve cell, i.e. by the so-called antidromic impulse. In the lower trace of B, but not C, there is a spike potential recorded as the impulse propagates down the axon of the nerve cell under observation. With rare exceptions, such as the climbing fiber synapses on Purkyně cells of the cerebellum, convergence of many presynaptic impulses is required to evoke an EPSP adequate for generating the discharge of an impulse from a nerve cell in the mammalian central nervous system. This is the principle of spatial summation which, as SHERRINGTON (1906) realized, is of fundamental significance in the integrative action of the central nervous system. Fig. 5 D–G shows that the generation of an action potential is an all-or-nothing event. There is no trace of it in D, which is a pure EPSP, but in E the EPSP reaches the critical size and the full-sized action potential is superimposed. Larger EPSP's in F and G attain this critical level earlier, so reducing the synaptic delay which may be defined as the time for transmission across a synapse. Normally this delay is less than one millisecond.

Fig. 5 H–K shows the operation of the other type of synapse, the inhibitory synapse. In H these synapses are seen to evoke a hyperpolarizing potential (the downward deflection), the inhibitory postsynaptic potential or IPSP. The antagonist action of the IPSP on the depolarization of an EPSP is shown in I. When appropriately timed and large enough, the IPSP can prevent an EPSP from attaining the critical level of depolarization for generating an action potential (compare J with K).

It is of great importance that there are these two kinds of action of nerve impulses at the contact regions or synapses. As illustrated in Fig. 5B and E–G a sufficient summation of the synaptic excitatory action causes a nerve cell to fire an impulse down its axon, so that it in turn hands on the information it has received to its own synaptic contacts with many other nerve cells. In this way messages can be transmitted seriatim to successive relays of nerve cells, so giving a progressively expanding excitation. The number of excitatory synapses on a nerve cell may be very large. For example on pyramidal cells such as E in Fig. 3A there would be more than 10,000 spine synapses (cf. Fig. 3 B, *a–d*), and all of these probably are excitatory (COLONNIER, 1968; SZENTÁGOTHAI, 1969; COLONNIER and ROSSIGNOL, 1969). It will be appreciated that the spreading excitation could indeed become explosive, leading to convulsive activity, if it were not for the inhibitory synaptic actions that control and limit it (cf. Fig. 5 J, K). This inhibitory synaptic action in the cerebral cortex has only recently become well recognized, and I would now suggest that inhibition has a significance in the functioning of the brain not inferior to that of excitation and that there may be as many nerve cells engaged in this inhibitory action as in the excitatory action. For example in Fig. 3A some of the small cells probably are inhibitory, and in Fig. 3B synapses e, f and g can be recognized as being of inhibitory type both by structure (cf. Fig. 4B) and by location. These two categories of nerve cells are quite distinct, as shown in Figs. 6, 7 and 8, where inhibitory nerve cells and their synapses are shown in black. In the mammalian brain there are no known examples of cells having an excitatory action by one set of their synapses and an inhibitory action by another set (ECCLES, 1969 b). It is impossible to illustrate the immense synaptic connectivity between nerve cells in the brain. Fig. 3A shows but 8 out of the many thousands which actually would be present, but unstained, in such a Golgi preparation of the cerebral cortex (cf. SHOLL, 1956; COLONNIER, 1966; SZENTÁGOTHAI, 1969).

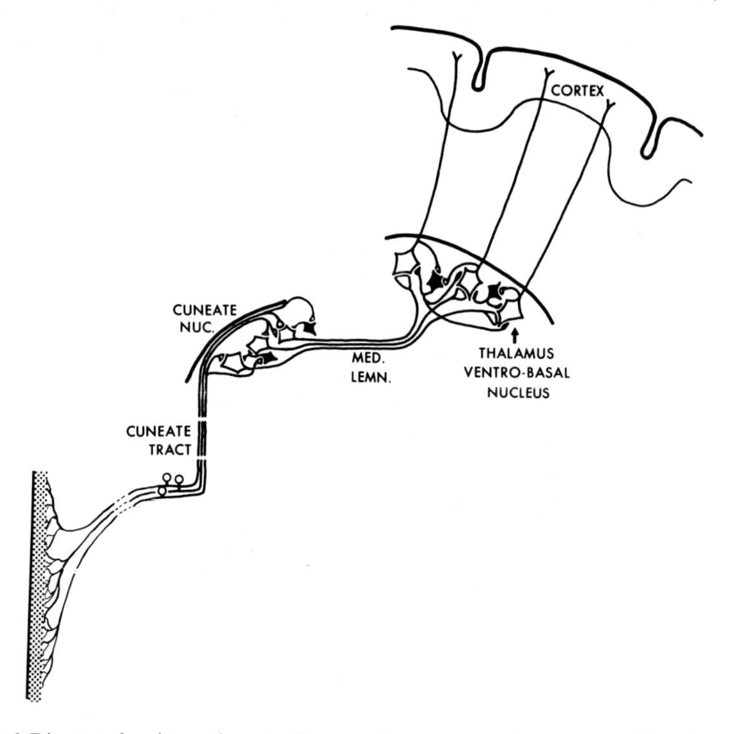

Fig. 6. Diagram showing pathway to the sensori-motor cortex for cutaneous fibers from the forelimb. Note the inhibitory cells shown in black in both the cuneate nucleus and the ventrobasal nucleus of the thalamus. The inhibitory pathway in the cuneate nucleus is of the feed-forward type and in the thalamus it is feed-back. By convention in this figure and the next two figures the inhibitory cells and their synaptic knobs are shown in black

A good illustration of transmission of information in the nervous system is provided by the pathway from sensory receptors in the skin to the somaesthetic area of the cerebral cortex. For example in Fig. 6 three nerve fibers from these receptors in the skin of the forelimb are shown passing up the spinal cord in the cuneate tract and so to synaptic relays in the cuneate nucleus. The pathway is then via the medial lemniscus, with synaptic relays in the ventrobasal nucleus of the thalamus and so by thalamo-cortical fibers to the cerebral cortex. At each relay station transmission is simple, being interrupted by only a single synapse, but there is also opportunity for inhibitory synaptic action by the small cells shown in black. Fig. 7 illustrates in a simplified form the complexities of neuronal operation encountered by the afferent impulses (the thalamo-cortical fibers of Fig. 6) reaching the cerebral cortex. Some of the branches of these afferent fibers give synapses directly to the pyramidal cells, and other branches go to excitatory and inhibitory stellate cells and so to pyramidal cells. The pyramidal cells in turn have axon collaterals, as

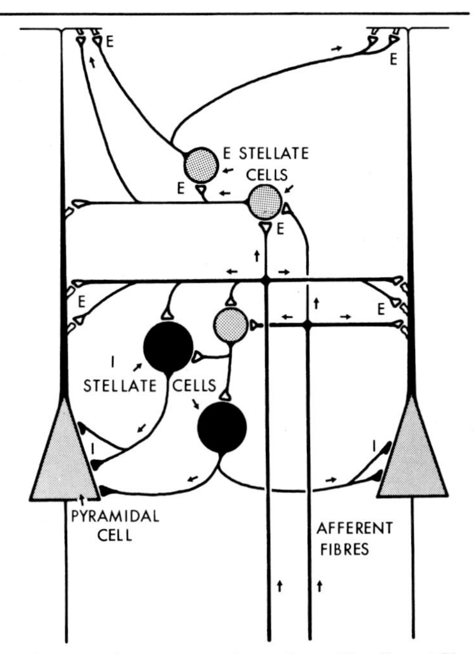

Fig. 7. Diagram showing postulated connections of specific afferent fibers (cf. Fig. 6) to the neocortical pyramidal cells. Note that the inhibitory path is through inhibitory cells that are activated either directly by the afferent fibers or by mediation of excitatory interneurones. Also note the various degrees of complexity of the excitatory pathways to the pyramidal cells. Arrows indicate direction of impulse propagation

shown in Fig. 8, that can result in an immense development of the excitatory input via the afferent fibers, there being again all possible interactions of excitatory and inhibitory nerve cells.

It is impossible to conceive the complexity actually obtaining in propagation over neuronal chains, where each neurone is linked to hundreds of other neurones and where the convergence of many impulses within a few milliseconds is necessary in order to evoke a discharge from any one neurone. It is therefore desirable to study the properties of simple models of neuronal networks as was done by Burns (1951, 1958), by Fessard (1954, 1961), by Cowan (1964), and more recently by numerous investigators.

Fig. 9 gives a diagrammatic representation of impulse propagation through a series of neurones arranged formally in the characteristic manner of in-parallel and in-series. This model is of course tremendously simplified, but it does help to illustrate the manner of operation of assemblages of nerve cells in the central nervous system. In the upper dia-

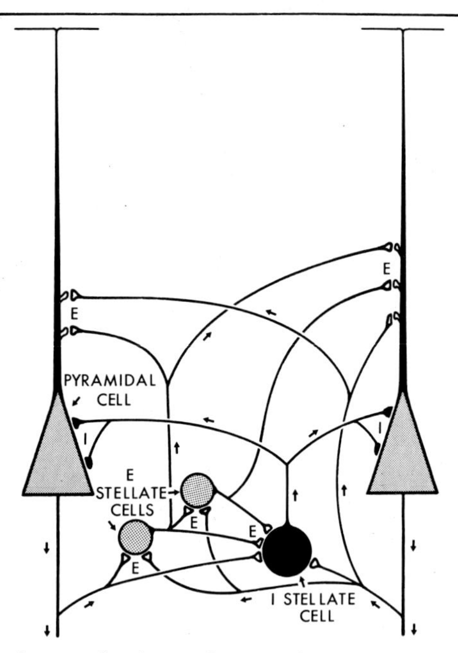

Fig. 8. Diagram showing postulated synaptic connections of axon collaterals of pyramidal cells. The single inhibitory interneurone together with its inhibitory synapses on the somata of the pyramidal cells are shown in black. All other stellate cells and the pyramidal cells are assumed to be excitatory and are shown open. The arrows indicate directions of impulse propagation. Note that, as suggested by the experimental evidence, both the excitatory and inhibitory pathways can include interpolated excitatory interneurones

gram two of the cells in column A are assumed to be excited and are in grey. Various axonal branches and their synaptic connections from cells in column A to column B and so on to C and D are shown. By convention in this model it is assumed that summation of two synaptic actions is required to generate the discharge of an impulse from a cell and so its activation is transmitted (see arrows) to synaptic terminals (grey knobs) on the cells of the next serial column. In this way it can be seen that the excitation of cells A 1 and A 2 will result in the output of impulse discharges from cells D 3 and D 4 but not from cells D 1 and D 2. As already mentioned, Fig. 9 is an extremely simplified model because, firstly, it assumes a simple geometrical arrangement with columns of neurones in serial array, whereas there are known to be all manner of synaptic connections bridging such serial order, as for example from cells of A column directly to cells of columns C and D. Also there are many varieties

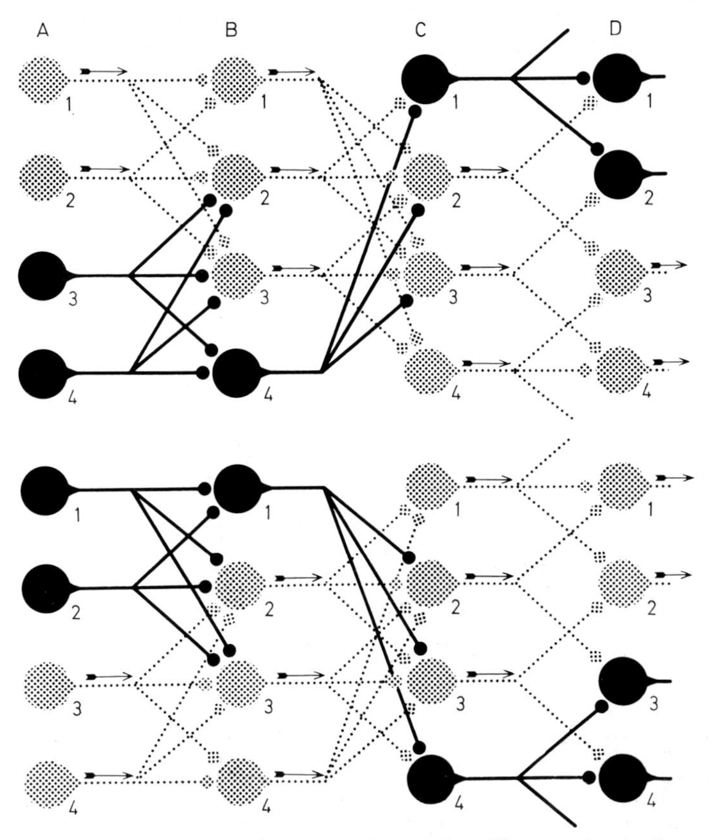

Fig. 9. Model of a highly schematic neuronal network to illustrate the simplest case of propagation along a multilane pathway. There is exactly the same anatomical network in the upper and lower diagrams. The synaptic connections of the twelve cells in columns A, B, and C are drawn, cells with impulses (note arrows) being shown light grey, while the silent cells are black. The assumption is that a cell fires an impulse if it is excited by two or more synapses (also light grey). Thus an input A_1A_2 results in an output discharge of D_3D_4 (upper diagram), whereas an input of A_3A_4 gives an output of D_1D_2 (lower diagram). Neurons B_2, B_3, C_2, C_3 are activated in the crossing zone for both these inputs. This diagram suffers from the serious defect that it ignores inhibition

of feedback controls via axon collaterals (cf. Figs. 3A, 8), and the consequent loop operation of large assemblages of cells. Another serious deficiency in Fig. 9 is that no account has been taken of inhibition. Nevertheless Fig. 9 is of value because it gives a simple diagrammatic illustration of the way in which information is propagated from neurone to neurone, summation being required at each synaptic relay. Consequently there is the necessity for the in-parallel arrangement of neurones.

The synaptic connections in the model of Fig. 9 from column A to B to C to D have been designed to illustrate one remarkable and important property of a neuronal network, namely, that two completely different inputs at A (A_1 A_2 in the upper picture or A_3 A_4 in the lower picture) can be transmitted through the same pattern of cell connections (A to B to C to D), crossing each other and emerging as completely different outputs at D (D_3 D_4 in the upper picture and D_1 D_2 in the lower). An interval of some milliseconds between the two wave fronts would eliminate interference by neuronal refractoriness or by summation of synaptic excitations. It will be noted that along the fringe of the advancing wave front in Fig. 9 there are subliminally excited neurones (C_1, D_1, D_2 in the upper picture). Such "fringe neurones" give opportunity for growth of a wave if other influences should also subliminally activate them.

Thus we are introduced to the concept of lability of a wave front. It may be diminished by inhibitory or depressant influences, and so ultimately be extinguished, or it may be enhanced by factors aiding in the activation of "fringe neurones." It will also be appreciated that, if a wave front moves into neuronal pathways having interconnections of suitable configuration, it may bifurcate into two waves propagating independently; while, conversely, two wave fronts propagating at the same time into the same pool of neurones would coalesce and give an onwardly propagating wave having features derived from both waves, with additional features due to the summation.

It will be realized that in the neuronal network of the cerebral cortex the factors involved in transmission of a wave front are vastly more complicated than in the simplified models of Fig. 9. In the first place, each neurone would make very many synaptic contacts with other neurones, and also receive from many more—probably from hundreds. Perhaps as many as one hundred neurones would be involved in each stage of an effectively advancing wave front and not 2 or 3 as in Fig. 9, so this wave would sweep over 100,000 neurones in one second. An advancing wave would also be branching at intervals, often abortively, and coalescing with other waves to give a complex spatio-temporal pattern.

Fig. 10 can be regarded as an enormous development from the simple array of Fig. 9. In order to illustrate patterned operation of hundreds of neurones there has been suppression of all diagrammatic representation of nerve fibers and synapses (cf. Fig. 9). In the diagram it is assumed that adjacent cells are synaptically connected in the forward direction indicated by the arrows. Inactive cells are shown in light grey. The dark grey and black arrangements of cells in serial order are meant to illustrate the propagation of impulses along chains of neurones arranged in parallel with branches and convergences. At each stage of serial relay there is assumed to be synaptic summation with the consequent discharge of some cells as in Fig. 9. It will be understood that this is an entirely imaginary diagram of a small fragment of a spatio-temporal pattern of neuronal activation and that it is constructed in order to convey some insight

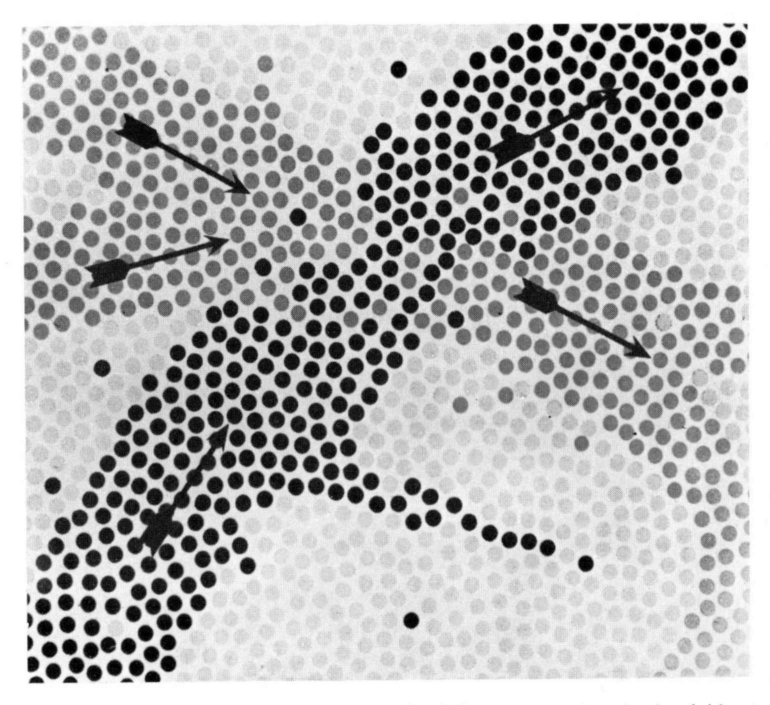

Fig. 10. In this schematic diagram the nerve cells of the cortex are imagined as laid out as dots in one plane. The multilane traffic in one, evolving a specific neuronal pattern, is shown in black, and in another as dark gray. The light gray cells are not activated by either pattern. Note that at the crossing of these two lanes the same nerve cells should participate in both and so each should be represented as a very dark gray dot, not just dark gray or black as shown. Arrows show direction of propagation (ECCLES, 1958)

into the extremely complex patterned operation of large neuronal assemblages such as occur in the cerebral cortex. It also illustrates the important property that any one nerve cell can participate in a large number of separate neuronal patterns. Its responses can belong to this or that pattern according to the ensemble of neurones that are activated with it, as will be appreciated from the two models in Fig. 9. There is now much evidence to indicate that the input of impulses via channels from receptor organs is itself repetitive and evokes long lasting serial activity in immense numbers of neurones linked up in as yet unknown pathways (FESSARD, 1961; further references in ECCLES, 1966b). Furthermore there is continuous on-going activity of the cortical pyramidal cells, and this activity can be raised or depressed by sensory inputs (POWELL and MOUNTCASTLE, 1959; EVARTS, 1961, 1962, 1964; HUBEL and WIESEL, 1962, 1963, 1965).

Neurophysiological Events Relation to Perception

We are now well informed in respect of the initial stages of the process whereby stimulation of receptor organs gives rise to a perceptual experience. We owe our understanding of the coding of such stimulation into trains of nerve impulses to the pioneer contributions of Adrian and his school; and the processes of transmission through the serial synapses on the pathway to the cerebral cortex (cf. Figs. 6, 7) have been illuminated by contributions to a recent symposium (MOUNTCASTLE, 1966a; GRANIT, 1966; CREUTZFELDT *et al.*, 1966). In this pathway there is much integration in the coded information, but this can be but a prelude to the unimaginably complex integrational procedures in the cerebral cortex that have been so effectively discussed by FESSARD (1961).

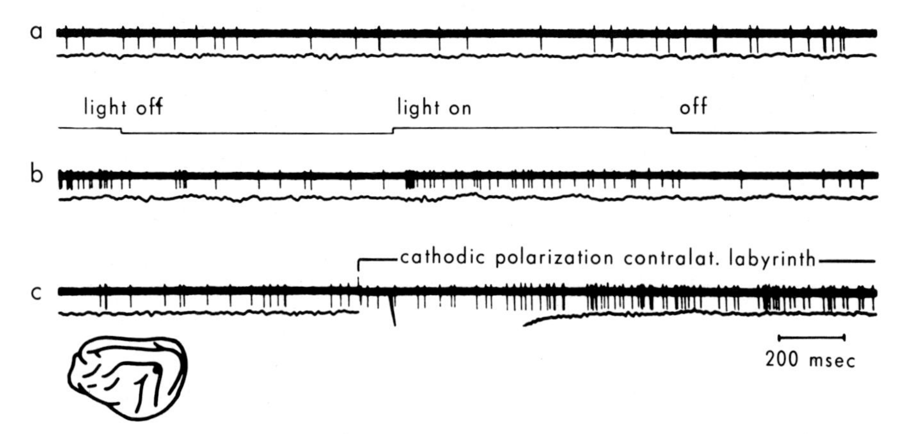

Fig. 11a–c. Extracellular responses of neurone in an association area of the neocortex (the para-auditory cortex as shown in inset diagram). a gives spontaneous activity in darkness; b shows that it responds to light and is transiently depressed by the onset of darkness; c shows that it is strongly excited from the contralateral labyrinth (JUNG, KORNHUBER, and DA FONSECA, 1963)

Indubitable evidence of the convergence of sensory information in the cortex is provided by those experiments in which a single cortical neurone is shown to be activated from several different sensory inputs. Multisensory convergence onto cortical neurones has been studied very intensively by JUNG and his collaborators (JUNG, 1961; JUNG, KORNHUBER, and DA FONSECA, 1963; KORNHUBER and ASCHOFF, 1964); and also by DUBNER and RUTLEDGE (1964). For example, in Fig. 11 a neurone

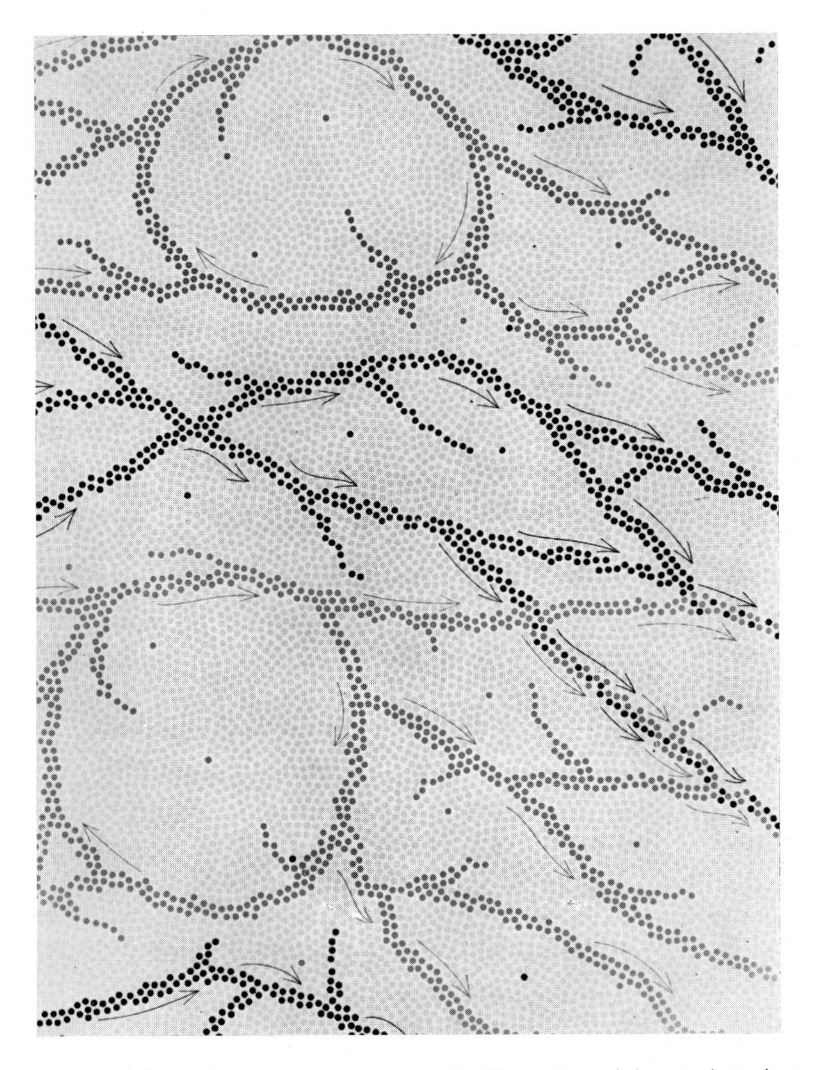

Fig. 12. By employing the same conventions as in Fig. 10, an attempt is here made to give a more elaborated representation of the enormous complexity of the spatiotemporal patterns of propagating activity in the cerebral cortex. The arrows show the directions of propagation for two separate evolving patterns (black and dark gray). Note that the black and gray patterns coalesce at two sites and propagate as one advancing wave. This would correspond to some integrative fusion of different sensory inputs to give some perceptual synthesis (ECCLES, 1958)

in the para-auditory cortex was excited both by light and by vestibular stimulation. Similar convergence of two or three sensory modalities was observed in all cortical areas investigated, both in the primary receiving areas and in the associative areas of the cortex. BUSER and IMBERT (1961) have obtained comparable results in an extensive study on multisensory neurones. I should also refer to the very comprehensive report of sensory integration in the cortex as signaled by evoked potentials and their interaction (ALBE-FESSARD and FESSARD, 1963). Provisional attempts have been made to discover the physiological significance of these convergence patterns, but for our present purpose it is sufficient to document this evidence that there are now hundreds of examples of multisensory convergence on single cortical neurones in all parts of the cerebral cortex, as well as in all the subcortical centers.

It is difficult to conceive the complexity actually obtaining in propagation over neuronal chains, where each neurone is linked to hundreds of other neurones and where the convergence of many impulses within a few milliseconds is necessary in order to evoke a discharge from any one neurone. Fig. 12 is an attempt to illustrate, by the same conventions as Fig. 10, the complex spatio-temporal patterns of neuronal activation evoked by two different sensory inputs that eventually come to converge and activate the same neurones. The convergent sites are indicated by arrows. The operational properties of Fig. 12 will be referred to in several later chapters.

Chapter III

Synaptic Mechanisms Possibly Concerned in Learning and Memory (cf. Eccles, 1966b, 1966c)

Introduction

The word "memory" is a term now being used for a very wide range of phenomena that involve the storage of information and the retrieval or read-out of this information. Hence we have genetic memory, as, for example, may be illustrated by the statement of Eigen (1964) that "DNA for instance has memory." Also the term is used in relationship to immunology (immunological memory). Such diversity of usage means that the term *memory* as such needs specification. I shall adhere to the original restricted sense, and associate memory with that property of the central nervous system whereby it is effective both in the storage and retrieval of information. This property has been termed *psychic memory*, and may be considered to have two components, learning and remembering, corresponding to storage and retrieval of information.

It is generally recognized that there are two varieties of psychic memory. First of all there are brief memories for seconds or minutes, many examples of which have recently been considered by Brown (1964). One example is the ability to repeat sequences of numbers that have been read out. After a few seconds or minutes this memory is lost beyond all recall. Following Hebb (1949), Gerard (1949), and Burns (1958), one can postulate that such a memory is subserved by a spatio-temporal pattern of propagated impulses in some sort of reverberatory circulation. So long as this specific pattern is preserved in dynamic operation, retrieval is possible. The spatio-temporal pattern forms a dynamic engram such as Lashley (1950) would have envisaged. The second kind of memory is distinguished by its enduring character, even for a lifetime, and has been shown in many experiments to survive, even when the central nervous system is reduced to a quiscent state, as by deep anaesthesia, coma, or extreme cooling. Such memories must therefore have as a basis some enduring change that is built into the fine structure of the nervous system, and that is often referred to as a "memory trace," so that we have what is called the "trace theory of memory" (GOMULICKI, 1953). There is an immense literature on what we may term the *learning process*, which includes the phenomena of conditioned reflexes. It is impossible here to review it, but reference can be made to MORRELL (1961b) and KANDEL and SPENCER (1968) for recent comprehensive surveys. I will confine my treatment to the possible synaptic mechanisms concerned in learning and recall.

As a neurologist, I would assume that in long-term memory the structural change is essentially synaptic, but before considering this postulate in detail, I should refer to some recent alternative suggestions that are built upon a supposed analogy between psychic memory and either genetic memory (HYDÉN, 1959, 1964, 1965, 1967) or immunological memory (SZILARD, 1964). I do not propose to consider SZILARD's theory of memory and recall because it is based on several unacceptable assumptions in regard to the mode of operation of synapses.

Molecular Memory

No doubt everybody is familiar with the great popularity that has been achieved by HYDÉN's theory, the so-called "molecular memory." The claim is repeatedly made by exponents of molecular memory, or in the more extensive field now called molecular neurology (SCHMITT, 1964), that the theories of conventional neurology have been most unfruitful and that they have neither sufficient precision in formulation nor sufficient experimental basis. In view of this criticism it is surprising to find that theories of molecular memory are remarkable for extreme extrapolation beyond the experimental evidence. By very elegant techniques, HYDÉN (1959, 1964, 1965, 1967) has actually shown that nerve cells involved in a variety of learning situations have an increased RNA content, and they also often show an increase in the ratio of purine to pyrimidine bases (actually of adenine to uracil). One can readily agree that activity of nerve cells is likely to be associated with an increased protein production, which, of course, is dependent on the increased RNA content; but this is very inadequate evidence for the elaborate theory that has been developed by HYDÉN and which involves assumptions that are contrary to a great deal of neurophysiological evidence.

The theory of molecular memory makes a series of postulates which may be listed in serial order (cf. HYDÉN, 1965, pp. 226–232). (1) In an acute learning situation there are specific time patterns of frequency responses set up in neurones. (2) In some way, not stated, any particular specific frequency pattern (also called a modulated frequency) causes

DNA to produce uniquely specific RNA. (3) This RNA synthesizes specific proteins in the soma of the neurone. (4) These proteins in turn give rise to the production of the synaptic transmitter substance. Essentially it is postulated that by these four sequential operations a particular modulated frequency of activation results in a chemical specification of neurones, so that (5) a similar temporal pattern at some later stage evokes by a resonance-like reaction an increased transmitter production. HYDÉN (1964) further develops this postulate: "A chemical specification of neurones in learning, involving synthesis of chromosomal RNA and specified proteins, reacting on modulated frequencies in millions of neurones in different parts of the brain, stronger in some, weaker in others, would also fit the conception that a complicated task is learned and remembered, not as a series of bits, but in whole contexts. More easily specified areas of the brain, for example, the hippocampus in mammals, would get a more dominant position in learning."

It is evident to neurophysiologists that this speculation derives from the postulate that the frequencies of impulse discharge by nerve cells carry an extraordinary specificity of coded information. The intensive study of many varieties of neurones in the central nervous system certainly shows wide varieties in their temporal patterns of impulse discharge, but it does not show sharp specificity either in the patterns or in the frequency of discharge. In particular, the frequencies of discharge vary over wide ranges dependent upon the intensity of activation. One must reject the postulate that specificity is carried by some frequency modulation in the manner required by HYDÉN's theory. The immense wealth of neuroanatomical and neurophysiological data provides the framework for all the specificity that is required from the brain, and there is no necessity whatever for the speculative suggestions of additional specificity dependent upon frequency modulation. There is the further unacceptable postulate in HYDÉN's theory of molecular memory (No. 5 above) that there is a kind of resonance phenomenon involved in the recall by a subsequent frequency modulation resembling that responsible for the initial chemical specification. Furthermore, the evidence for the immense number of differently specified RNA molecules, one for each memory, derives merely from a significant change in the ratio of purine to pyrimidine bases, which could signify only two RNA's, yet thousands of millions are assumed to be specified. Later we shall see that HYDÉN's elegant demonstration of RNA increase in a learning situation can be built into the classical growth theory of learning; and, in fact, must be a necessary postulate of this theory, though of course there is no requirement of such a high order of chemical specification of this RNA. It should be stated that HYDÉN (1967) has since modified his hypothesis and eliminated some of the less acceptable features.

Synaptic Properties and Memory

The neuronal mechanisms concerned in the operation of the nervous system are of particular significance in relation to the development of explanations of the phenomena of learning, which may be demonstrated by some behavioural study. It is recognized that learned behaviour signals changes that have occurred in the neuronal connectivity within the central nervous system. We have seen in Chapter II that communication between the individual units of the nervous system (neurones) occurs at highly specialized regions of very close contact, the synapses. It is generally accepted now that the changes in transmission between neurones result from changes in the efficacy of the synapses (cf. ECCLES, 1964, Chapters 6, 7 and 16). Necessarily, the postulated changes in synaptic efficacy must be of very long duration—days or weeks. There is no way in which relatively brief durations of synaptic change for each synapse of a serial arrangement can sum to give a more prolonged change.

Implicit in most anatomical and physiological theories of learning is the concept that synaptic activation leads to an increased effectiveness of the synapse, and that, with a sufficient repetition of this activation, there is a prolonged stabilization of this increment (RAMÓN Y CAJAL, 1911; HEBB, 1949; TÖNNIES, 1949; YOUNG, 1951a; ECCLES, 1953, 1961, 1966b; KANDEL and SPENCER, 1968). There is further the complementary concept that diminished usage leads to regression in effectiveness, and thus an explanation of forgetting is to hand.

Following LASHLEY (1950) it is accepted that the learning of even the simplest behavioural response involves a complex patterned operation in millions of neurones, the so-called engram. Figs. 10 and 12 are attempts at diagrammatic display of fragments of two such engrams composed of the black and the dark grey cells respectively. The increased efficacy of the synapses linking such patterned organizations of neurones ensures that, in response to a given input from receptors, this neuronal pattern is replayed in the brain in much the same spatio-temporal form each time that the learned behavioural response is evoked. In fact the behaviour arises from the muscular contractions induced by this neuronal activation. Thus we have to envisage that learning involves not simply changes in a few synapses but that there is potentiation of the synapses linking immense assemblages of neurones, so providing the background conditions for spatio-temporal patterns of neuronal activity such as those depicted in Figs. 10 and 12. Only on the basis of such a concept can the learning of behavioural responses be explained.

Let us now consider properties of some synapses that I think are of great functional significance and that may be especially related to the learning mechanism.

Frequency Potentiation

Fig. 13A, B gives examples of intracellular records of EPSP's from neurones that are activated monosynaptically at various frequencies. It shows the surprising result that different synapses, even on the same cell, vary

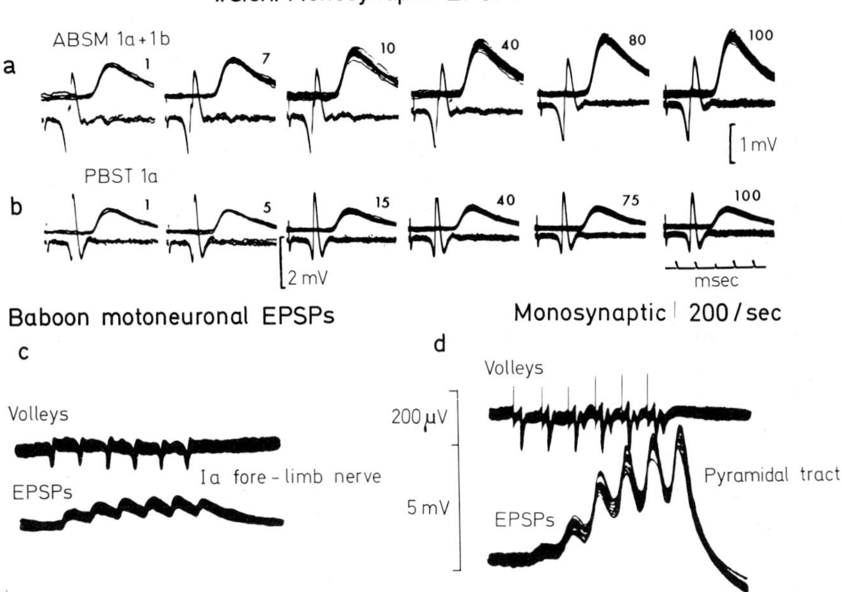

Fig. 13A–D. The effect of repetitive activation on sizes of monosynaptically produced EPSP's. A and B are superimposed traces of intracellular EPSP's evoked in a cell of the ventral spino-cerebellar tract by afferent volleys from muscle nerves as indicated—ABSM, anterior biceps semimembranosus; PBST, posterior biceps semitendinosus. The traces were photographed at the steady state reached after the first few stimuli at the frequencies per second indicated for each trace. The lower traces show the afferent volleys entering the spinal cord via the L7 dorsal root (ECCLES, HUBBARD and OSCARSSON, 1961). The lower records of C and D are superimposed traces of monosynaptic EPSP's evoked in the same motoneurone of the cervical enlargement of the baboon spinal cord. In each case there were six stimuli at 200/sec, which were to the muscle afferents in C and to the pyramidal tract in D. The upper traces are monitored presynaptic records (LANDGREN, PHILLIPS and PORTER, 1962)

greatly in their response to rapid repetitive stimulation. For example, the EPSP's (cf. Fig. 5) generated by the synapses in Fig. 13A double in size as the frequency is raised from 1/sec to 7, 10, and so on, up to 100/sec. By contrast on the same neurone (B) there is a slight decline at the highest frequencies for the EPSP's generated by other presynaptic fibers. Evidently these highest frequencies may impose quite a strain on the chemical transmitting mechanism. When we consider what is happening at a synapse, we can recognize that complicated mechanisms are necessary for ensuring effective high-frequency activation. The synaptic vesicles have to migrate up to the "firing line" fronting the synaptic cleft (cf. Figs. 3B, 4A, B, D) and then be discharged of their contents and replenished. Meanwhile they have to be replaced by other vesicles, which may have to be manufactured. Yet, all this can happen at a frequency of 100/sec, so that there may be little falling off in the amount of transmitter liberated by each successive impulse (Fig. 13B), or even a large potentiation as in Fig. 13A. Again, in Fig. 13C this very effective maintenance occurs with the monosynaptic pathway to motoneurons (cf. Fig. 5B–G).

Fig. 13D shows that synapses formed by cerebral neurons are more powerfully potentiated by high frequency stimulation than are spinal synapses (LANDGREN, PHILLIPS and PORTER, 1962). The intracellular records of Fig. 13C, D are from the same motoneurone in the cervical spinal cord of a baboon. In D the excitatory postsynaptic potentials (EPSP's) are monosynaptically generated by descending volleys of the pyramidal tract excited at 200/sec. In the series of Fig. 13C, this same motoneurone is monosynaptically excited by synapses coming from peripheral stretch receptors of muscle. The first EPSP in D has about the same size as that evoked from the muscle afferent nerve, but in successive responses it builds up enormously, by a factor of 10, and it tends to stabilize at this greatly potentiated level. This shows that synapses made by pyramidal tract cells have an extraordinarily efficient mechanism for increasing their effectiveness when they are stressed by high frequency stimulation. PORTER (1970) has shown that even at much lower frequencies of activation there is a significant potentiation of these pyramidal tract synapses, and hence suggests that it is important at physiological frequencies of activation.

This performance does not show, of course, that these cerebral synapses have learned anything. But I would suggest that when synapses have this enormous plasticity, as expressed by their increase in potency with high frequency stimulation, some residual potency is likely to remain, and this could be the basis for a learning theory. Frequency potentiation is one of those properties which suggests that cerebral synapses tend to be different from lower-level ones, and this may be linked with the greater learning ability which is exhibited in behavioural and conditioning experiments.

Fig. 14 gives an example of frequency potentiation for synapses on the granule cells of the hippocampus that are activated by stimulation of the entorhinal cortex (LØMO, 1970). The intracellular records at 1/sec show a very small initial excitatory synaptic action followed by a large inhibitory action (cf. Fig. 5H). With a stimulus at 10/sec, already within one second there is a large potentiation of the excitation that counteracts to some extent the inhibition. After 3 seconds of this stimulation the excitation is seen to generate two impulses in the cell, which appear as the sharp downward deflections in the subjacent extracellular record. On again slowing the stimulation to 1/sec, the frequency potentiation has already considerably declined at 0.4 seconds and has disappeared in

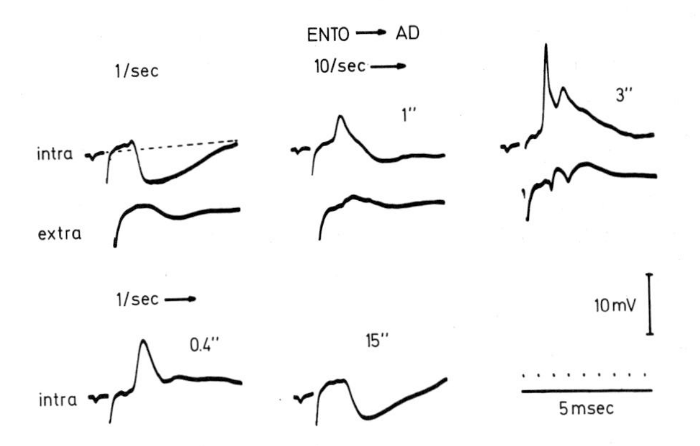

Fig. 14. Frequency potentiation of granule cells of fascia dentata of hippocampus. Entorhinal region was being stimulated initially at 1/sec, then for several seconds at 10/sec, and finally in the lower row at 1/sec again. Intracellular records (intra) are from a granule cell, while in the upper three records there are also extracellular records (extra). The voltage scale obtains only for the intracellular records. Further description in text (LØMO, 1970)

15 seconds. Much more investigation of such precisely defined monosynaptic pathways in the cerebral cortex is evidently needed. This work of LØMO (1970) confirms earlier reports by ANDERSEN, HOLMQVIST and VOORHOEVE (1966).

Post-Tetanic Potentiation

Besides this frequency potentiation during repetitive stimulation, there is an additional process of potentiation that follows after high frequency activation of synapses and continues often for quite a long time. Ever since this phenomenon of post-tetanic potentiation was discovered by LLOYD (1949) it has been regarded as offering a possible paradigm of the process of learning at synapses (ECCLES and MCINTYRE, 1953; CURTIS and ECCLES, 1960; KANDEL and SPENCER, 1968).

Fig. 15 (BLISS and LØMO, 1970) shows that the synapses for the entorhinal pathway onto dentate granule cells of the hippocampus are very strongly potentiated by conditioning tetanization and this potentiation can be cumulatively built up and then survives for many hours in a fully developed form. In fact there was no decline from the strongly potentiated level during the whole period of the experiment. The first five arrows in Fig. 15 signal brief conditioning tetani (20/sec for 15 sec), each one of

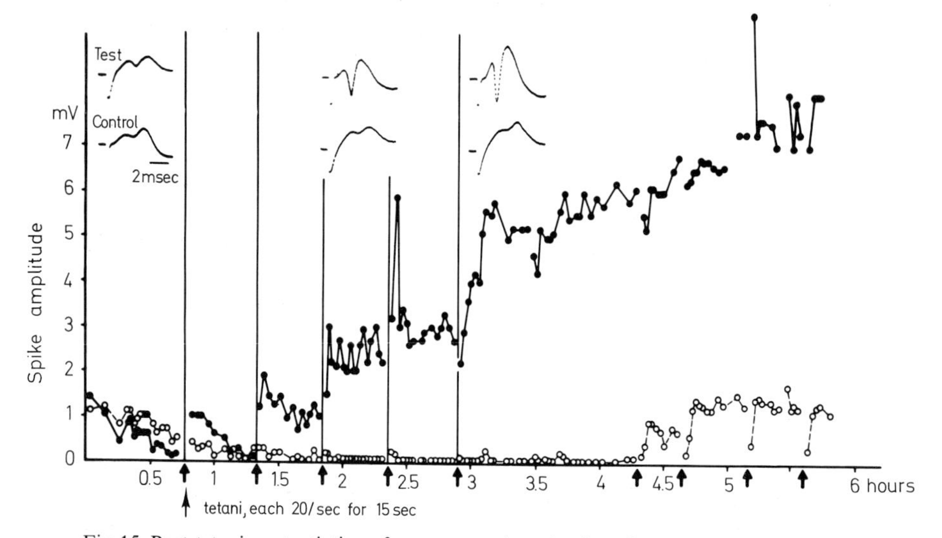

Fig. 15. Post-tetanic potentiation of monosynaptic activation of granule cells in response to entorhinal stimulation. The impulse discharge evoked by granule cells is signalled by a "population spike" recorded in specimen records as a brief downward (negative) deflection occurring during the slow positive wave. Five conditioning entorhinal stimuli (20/sec for 15 sec) were applied at the times of the vertical lines (see arrows below base line). Three averaged specimen records are shown for the testing response, one before, one after the third conditioning and one after the fifth conditioning. The circles plot the sizes of the observed spike potentials in mV, the solid circles being for the side that was subjected to the initial conditioning. The contralateral side was similarly conditioned by entorhinal tetanization at the times of the four later arrows, and a prolonged, though small potentiation is shown (BLISS and LØMO, 1970)

which is followed by a potentiation of the impulse discharges as shown by the sharp downward deflections of the three upper inset records. This increase in number of granule cells discharging impulses grows in step-wise manner to the extremely large potentiation after the last tetanus. The open circles show the control responses on the unconditioned side, which meanwhile exhibits no potentiation. However, at the four arrows late in the series, that side was stimulated and a small prolonged potentiation was induced. This experiment of BLISS and LØMO is very encouraging indeed for the proponents of the "synaptic usage" theory of learning because it shows that synapses in a part of the central nervous system that is believed to be importantly concerned in memory have a very highly developed mechanism for potentiation as a consequence of excess stimulation. Furthermore it shows the urgency for much more of these well designed investigations at higher levels of the nervous system.

The synapses involved in these experiments of LØMO are the spine synapses on the dendrites of granule cells, which would correspond approximately to those shown in Figs. 3B a, b, c and 18A. It is now generally recognized that the spine synapses are almost uniformly excitatory in character, having round vesicles (COLONNIER, 1968); and furthermore there is now much evidence to show that spine synapses are plastic in the sense that they exhibit regression and development to a quite remarkable degree under appropriate experimental conditions. For example in Fig. 18B there are several examples of secondary spines (SS) that bud off from the dendritic spines of Purkyně cells in the cerebellum, and each of these secondary spines appears to participate in a synapse.

Synaptic Regression with Disuse

There are now beautiful pictures to show the regression of spine synapses occurring under experimentally imposed conditions of disuse. For example, in Fig. 16A and C (VALVERDE, 1968) there are shown dendrites of the normal visual cortex of mice 24 and 48 days old respectively. By contrast, B and D show the very depleted spine population along comparable dendrites of the visual cortex contralateral to that for specimens A and C respectively. These animals had suffered enucleation of one eye at birth and hence the pyramidal cells of the contralateral visual cortex in specimens B and D had not been stimulated by sensory inputs. The striking depletion of spines in these two examples shows that, at least in the early periods of life, natural stimulation is of the greatest importance for maintaining and developing the dendritic spines. Evidently this essential trophic influence is attributable to the synaptic excitatory input onto pyramidal cells. This experiment on experimentally induced disuse is the obverse of the learning experiment. However, in general it supports the hypothesis that learning is due to usage because it demonstrates the regression that results from disuse. Fig. 17 gives a graphic illustration of disuse produced by raising animals in the darkness and having their eyes intact (VALVERDE, 1967; RUIZ-MARCOS and VALVERDE, 1969). In A the dendritic spines of the pyramidal cells of the visual cortex are seen to be depleted (filled circles) relative to control mice raised under normal conditions (open circles). On the other hand Fig. 17B shows that in a sample of non-visual cortex the two populations of mice had approximately the same density of spines on their pyramidal cell dendrites.

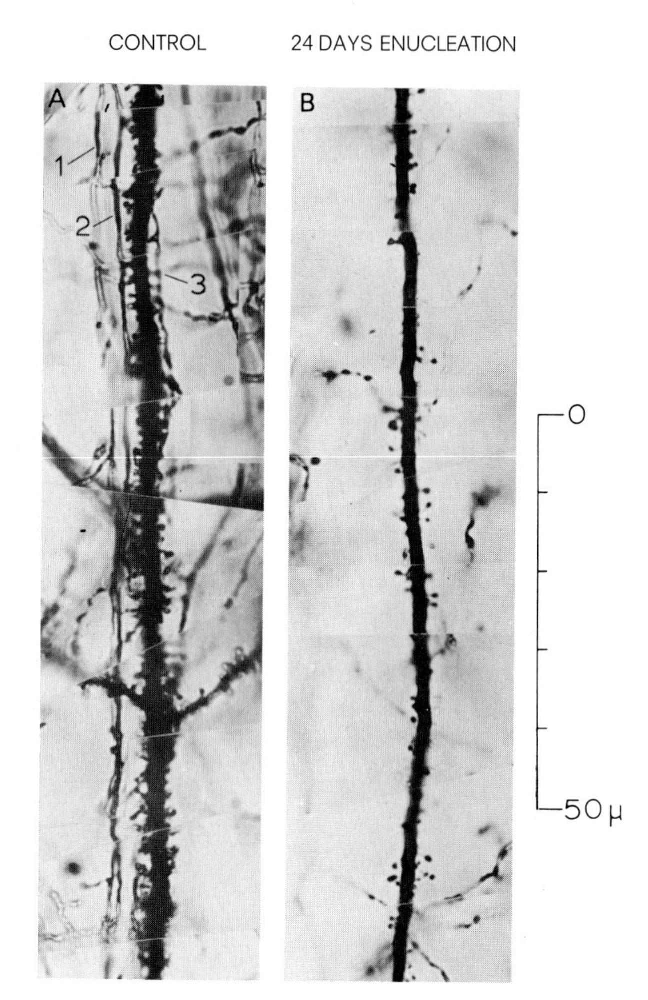

Fig. 16 A–D. Mosaic microphotographs of apical dendrites of pyramidal cells in the mouse area striata. A and B are specimen pictures of Golgi stained cortex from a mouse 24 days old that had one eye enucleated at birth, A being ipsilateral and B contralateral to the

The experiments of VALVERDE recall those of WIESEL and HUBEL (1963 a, b) in which there were histological and physiological investigations of nerve cells in the visual pathways of kittens with one eye deprived of normal vision from birth. They found that this deprivation led to unresponsiveness of cells in the visual cortex to stimulation of the deprived eye. It appears that the connections already found to be established at birth had regressed. When the visual deprivation was applied later in

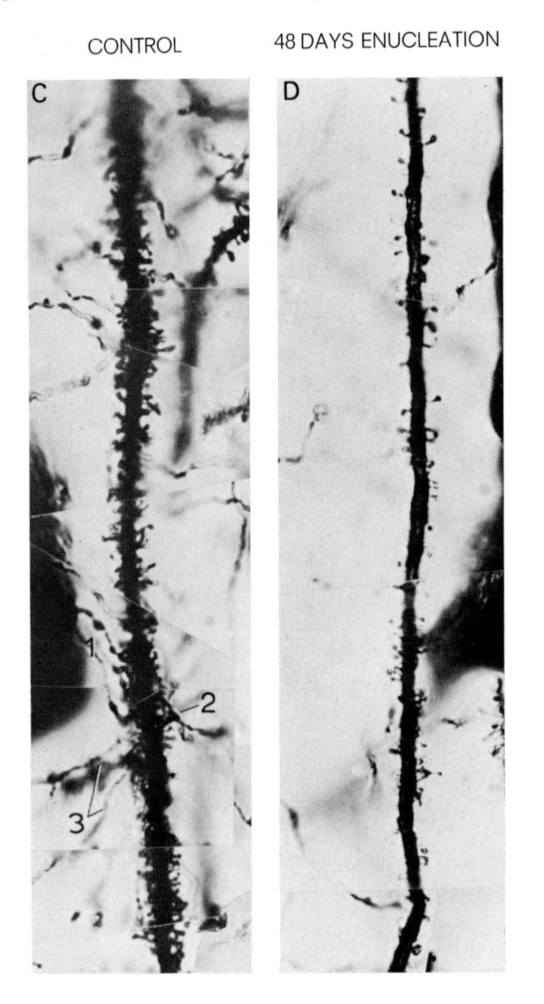

enucleation. Note the parallel fibers 1, 2 and 3 escorting the dendrite in A. C and D resemble A and B except that mouse was 48 days old (VALVERDE, 1968)

life in visually experienced kittens, little if any physiological abnormality was induced. Evidently the plastic properties of the visual pathways are much more in evidence in the first few weeks of life. This of course is in line with VALVERDE's experiments, which were done on very young mice.

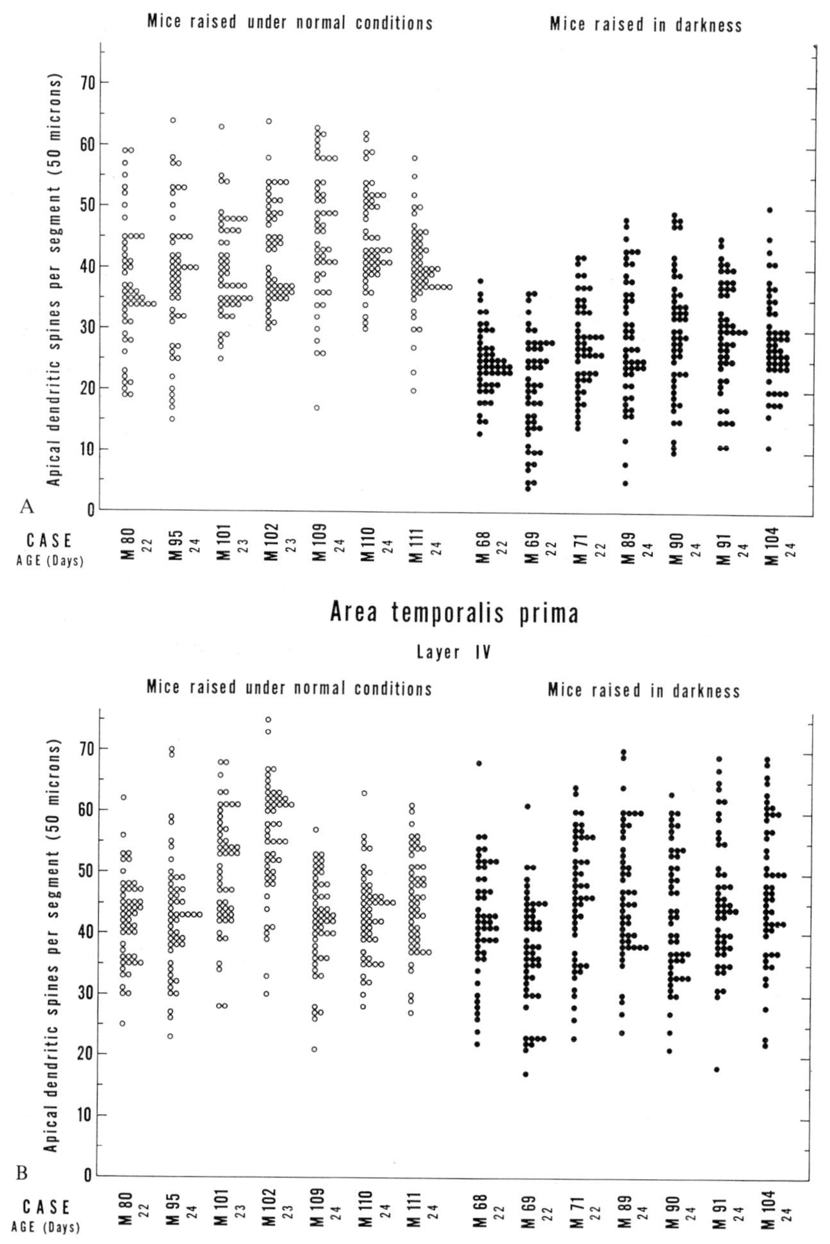

Area striata
Layer IV
Mice raised under normal conditions
Mice raised in darkness
Apical dendritic spines per segment (50 microns)
A
CASE
AGE (Days)
M 80 22
M 95 24
M 101 23
M 102 23
M 109 24
M 110 24
M 111 24
M 68 22
M 69 22
M 71 22
M 89 24
M 90 24
M 91 24
M 104 24
Area temporalis prima
Layer IV
Mice raised under normal conditions
Mice raised in darkness
Apical dendritic spines per segment (50 microns)
B
CASE
AGE (Days)
M 80 22
M 95 24
M 101 23
M 102 23
M 109 24
M 110 24
M 111 24
M 68 22
M 69 22
M 71 22
M 89 24
M 90 24
M 91 24
M 104 24

Fig. 17

Discussion of Growth Theory of Learning

The simple concept that disuse leads to regression of spine synapses (cf. Fig. 18) and excess usage to hypertrophy can be criticized because it is now recognized that almost all cells at the highest levels of the nervous system are discharging continuously. One can imagine therefore that there would be overall hypertrophy of all synapses under such conditions, and hence no possibility of any selective hypertrophic change. Evidently, frequent synaptic excitation could hardly provide a satisfactory explanation of synaptic changes involved in learning. Such "learned" synapses would be too ubiquitous. This criticism is of particular interest in relationship to some recent suggestions that have been proposed by SZENTÁGOTHAI (1968) and MARR (1969) in relation to cerebellar learning, and which, adopting MARR's terminology, we may call the "conjunction theory" of learning. This proposal is particularly attractive because it makes sense of the otherwise inexplicable overlap of the two completely different systems of input to the cerebellum, namely by mossy fibers and climbing fibers (cf. ECCLES, ITO and SZENTÁGOTHAI, 1967). The mossy fiber input acts via granule cells and parallel fibers to excite Purkyně cells by spine synapses on their dendrites. According to the conjunction theory, there is no aftermath from such synaptic activation of Purkyně cells in itself. However, when there is conjunction of this excitation of the spine synapses with the massive excitation of the Purkyně cell provided by a climbing fiber, the conjunction theory postulates that the spine synapses retain thereafter some increased efficacy. It can be envisaged that this effect will grow and become more enduring with repeated conjunctions and in this way some pattern of mossy fiber input can become a strong and selective excitant for the Purkyně cells so influenced.

The conjunction theory of learning need not be restricted to the cerebellum, because it is now recognized that the pyramidal cells of the cerebral cortex also have fibers running up their apical dendrites, much as for the climbing fibers of the Purkyně cells (cf. Fig. 16A, 1, 2, 3; COLONNIER, 1966; VALVERDE, 1968; SZENTÁGOTHAI, 1969). However, the manner of activation of these cerebral "climbing fibers" is not yet understood.

Fig. 17A and B. Numbers of spines per 50 μ lengths of apical dendrites of mouse cortical pyramidal cells. In A open circles represent observations on the area striata of seven normal mice, and filled circles on seven mice raised in darkness. For each mouse specified below there were spine counts on 50 segments of apical dendrites, each 50 μ long and the counts are plotted in the respective columns. B represents control investigation in which similar counts were made on the same mice, but for pyramidal cell dendrites in the area temporalis prima (VALVERDE, 1967)

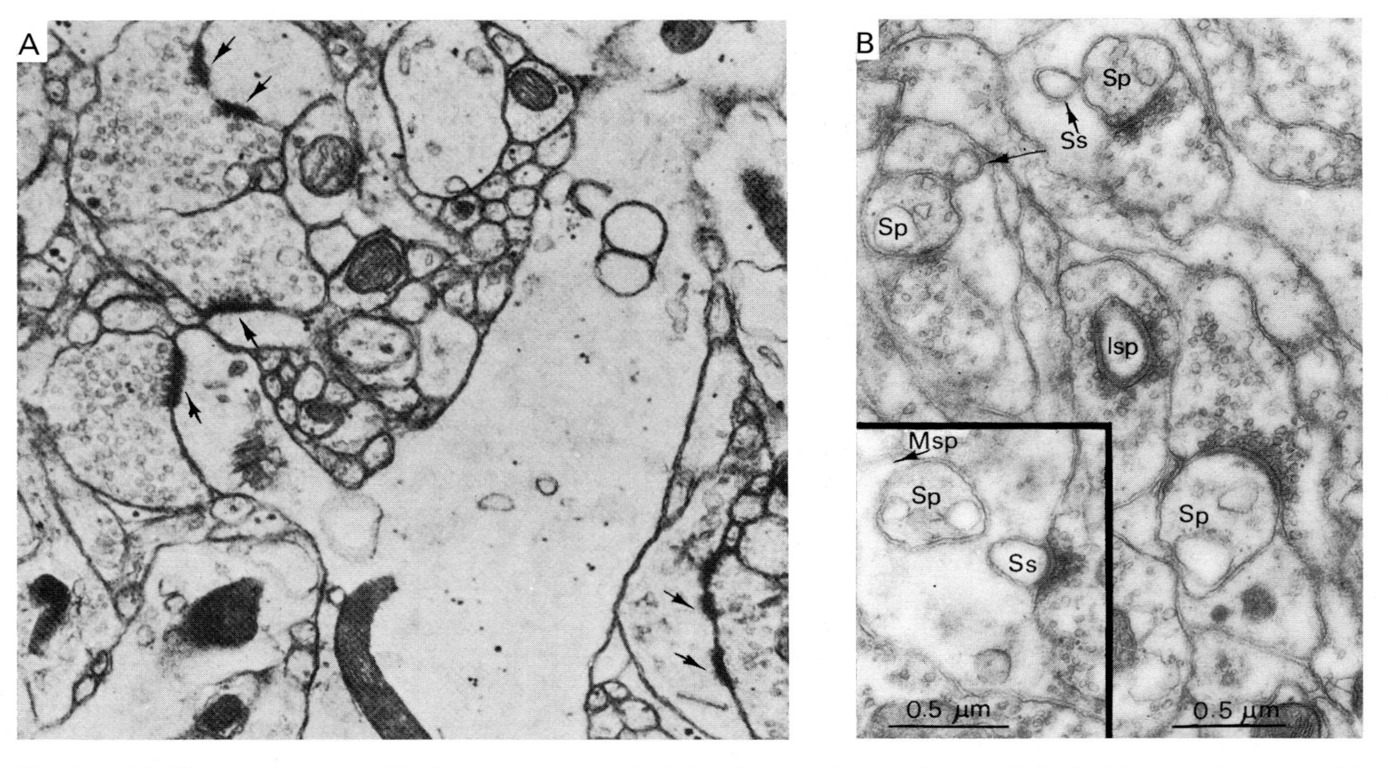

Fig. 18A and B. Electronmicrographs of dendritic spine synapses. In A there is a large dendrite of a pyramidal cell of the cat cerebral cortex with a spine that has on it a synaptic ending with round vesicles and the characteristic dense area (arrow) of the active synaptic site. Note the spine apparatus midway along the spine. Other spines with synapses are also indicated by arrows (COLONNIER, 1968). In B there are dendritic spines (*Sp*) of cerebellar Purkyně cells with presynaptic contacts by parallel fibers characteristically filled with synaptic vesicles. *Ss* show secondary spines budding off from the dendritic spines and *Isp* is a dendritic spine cut so that it appears to be embedded in a parallel fiber (HÁMORI and SZENTÁGOTHAI, 1964)

Biochemical Mechanisms in Synaptic Growth

Any attempt to explain learning by relating synaptic growth to activation must recognize that the operation of appropriate metabolic machinery is required to produce the postulated structural changes. It is pertinent to recall the evidence of HYDÉN (1964) and MORRELL (1961a, 1969) that there is an increase in the RNA content of neurones subjected to excess stimulation. Presumably, in the synaptic growth theory of learning, it must be postulated that RNA is responsible for the protein synthesis required for growth (cf. ECCLES, 1966b; STENT, 1967). However, this postulated growth would not be the highly specific chemical phenomenon postulated in HYDÉN'S molecular theory of learning. The specificities are encoded in the structure and in the synaptic connections of the nerve cells, which are arranged in the unimaginably complex pattern that has already been formed in development. From then onward, all that seems to be required for the functional development that we call learning is merely the microgrowth of synaptic connections already in existence, e.g. of the spine synapses on pyramidal cells and Purkyně cells (ECCLES, 1966b). The flow of specific information from receptor organs into the nervous system (cf. Figs. 6, 7 and 8) will result in the activation of specific spatio-temporal patterns of impulse discharges. The synapses so activated will grow to an increased effectiveness; hence, the more a particular spatio-temporal pattern of impulses is replayed in the cortex, the more effective become its synapses relative to others. And by virtue of this synaptic efficacy, later similar sensory inputs will tend to traverse these same neuronal pathways and so evoke the same responses, both overt and psychic, as the original input.

DINGMAN and SPORN (1964) have reviewed several experiments that indicate a relationship between the RNA content of the nervous system and learning ability. First, MORRELL (1961a) finds an increase in the RNA of nerve cells forming an epileptogenic mirror focus in the cerebral cortex, which may be regarded as a neurophysiological model of memory. Second, DINGMAN and SPORN (1964) find that a decrease of functional RNA depresses the rat's ability to learn a new maze, and CHAMBERLAIN, HALICK and GERARD (1963) report that under the same conditions, "fixation of experience" requires a longer time, whereas an increased RNA concentration shortens the time for fixation. There have been several reports that long-term administration of yeast RNA improves memory function but the mode of action is unknown (cf. QUARTON, 1967). There are now many reports of experiments on memory transfer by brain extracts alleged to be acting by RNA, but probably the effects are non-specifically evoked (cf. QUARTON, 1967).

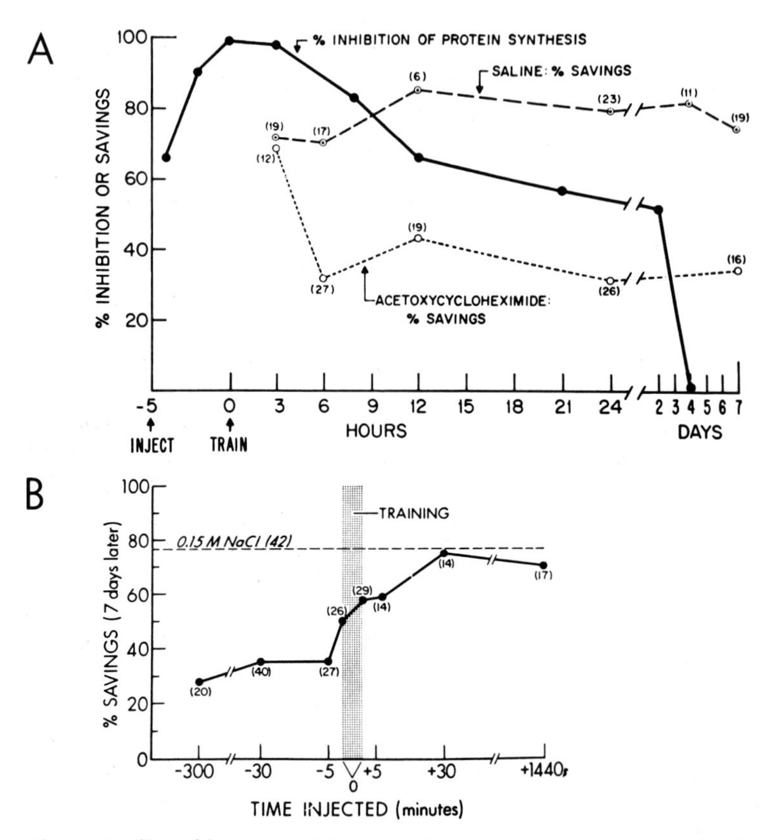

Fig. 19. A Effect of intracerebral injection of acetoxycycloheximide on cerebral protein synthesis and memory. Mice were injected intracerebrally with 20 μg acetoxycycloheximide 5 hr before training to escape shock by choosing the left limb of a T maze to a criterion of 3 out of 4 consecutive correct responses. Groups were tested for retention at the indicated time after training. Number of mice in each group indicated in parentheses (BARONDES and COHEN, 1967). B Effect on memory, of subcutaneous administration of acetoxycycloheximide at times before and after training. Mice were injected subcutaneously with 240 μg of acetoxycycloheximide at indicated time relative to training. Inhibition of about 90 percent of cerebral protein synthesis was established within about 10 min after injection. Training, designated as 0, took an average of 8 min. Mice were tested for retention 7 days after training. Numbers in parentheses are number of mice in each group. Mice injected with acetoxycycloheximide before or within 5 min after training all had significantly less savings than saline controls. Mice injected 5 or more minutes before training had significantly less savings than those injected immediately after training. Injections of acetoxycycloheximide 30 min or more after training had no significant effect on memory (BARONDES and COHEN, 1968)

An experimental investigation directly related to the growth theory of memory has been reported by BARONDES on mice (1969). Essentially the same results have been obtained by AGRANOFF on goldfish (1967). In mice, cerebral protein synthesis is depressed to about 5% of its normal level by intracerebral or subcutaneous injection of cyclohexamide or acetoxycycloheximide. After mice were injected, they exhibited as good a trained performance as control mice when tested up to 3 hours after the training period, but after 6 or more hours they displayed a large memory loss. Apparently the good performance up to 3 hours was due to short-term memory, but long term memory was greatly impaired (Fig. 19 A). In a second type of experiment (Fig. 19 B) it was found that there was no defect in long term memory when the injection was made at least 30 minutes after the training period, and that, if the injection was before or up to 5 minutes after the training period, there was a grave deficiency in long term memory. These results indicate that protein synthesis is required for long term memory and that the synthesis is virtually complete within 30 minutes after the training period.

Furthermore, on the basis of the experimental evidence of AGRANOFF (1967) on the goldfish, it is suggested by BARONDES that neuronal activation firstly leads to RNA synthesis (cf. DROZ and BARONDES, 1969) and this in turn to the protein synthesis (cf. GLASSMAN, 1969). This work provides excellent experimental support for the growth theory of learning. It further gives rise to the suggestion that the protein synthesis may be directed by newly synthesized messenger RNA or ribosomal RNA, which would be in accord with HYDÉN'S (1959, 1964, 1967) experimental evidence, as already mentioned. A correlation may be made with the evidence of MORRELL (1961a, 1969) that there is an increase in RNA in the mirror focus of an epileptic lesion of the cerebral cortex, which presumably arises because of the prolonged bombardment by impulses transmitted from the epileptic lesion via the corpus callosum.

The Engram and its Readout

We have seen that a long term memory must depend on enduring increase in the synaptic efficacy that has been built up in a specific neuronal pattern, and that is produced by an initial sensory input. One may surmise with HEBB (1949) that a significant synaptic change requires a reverberatory circulation of impulses many times through a specific neuronal pattern. In this connection it is relevant that an experience may not be remembered if a cerebral trauma (concussion or electric shock) is applied as long as 20 minutes later. However, only a part of this retrograde amnesia can be attributed to cessation of circulating impulses, for it is much briefer when cerebral activity is blocked by a rapid anaesthesia. One may surmise that the growth of synapses continues for some time after the circulation of impulses ceases, and that the cerebral trauma can block this growth process.

On account of the enduring increase in synaptic efficacy there will be a tendency for a particular pattern of neuronal activity to be evoked by a particular predisposing or triggering neuronal activity and/or excitatory afferent input. We may say that the remembered image is experienced while its specific spatio-temporal pattern is being replayed in the cortex. It may therefore be postulated that the initial development of effectiveness in certain synaptic junctions is a sequel to the primal event that is remembered, and this effectiveness is sustained and even enhanced by each subsequent replaying in the brain (and remembering in the mind).

The physiological equivalent to memory is the conditioned reflex, which has been shown by GASTAUT (1958) and JASPER, RICCI and DOANE (1958) to produce characteristic changes in the electroencephalogram. These changes are often observed over large areas of the cortex and may be assumed to be due to spatio-temporal patterns of neuronal activation that are spreading widely in the cortex along specific paths laid down during the conditioning.

These postulates of engrams accord well with one's experiences of remembered imagery. By far the best evocative situation arises from some closely similar experience. Here the evolving spatio-temporal patterns will tend to correspond closely to the original "congealed" pattern; hence the stage is set for the entry into and the replaying of this old pattern, as is shown for example in Fig. 12 for the grey, but not for the black pattern. In less favourable situations we may try to recall some imagery by various devices or tricks of memory, deliberately choosing specific sensory inputs or trains of thought for this purpose. How often do we attempt these techniques for remembering and how skilled do we become at it! Consider for example the attempts to recall a tune, or a visual image or a name or an event. We are all aware of the potency of particular predisposing conditions. Language whether spoken or written is overwhelmingly important in these respects, and becomes increasingly so with education and cultural development. Thus we learn to experience vicariously the imagery of writers and artists. Poetry is a particularly effective medium for the transmission of imagery, transcending time and place and appealing to all who have educated themselves to have in their cortex engrams that are in harmony with the cortical patterns evoked anew by reading some "pregnant" lines of poetry or better still by hearing them. Our use of the word "pregnant" is significant of our experience of the wealth of evoked imagery.

Summary

We may summarize this discussion of the structural basis of memory by stating that memory of any particular event is dependent on a specific reorganization of neuronal associations (the engram) in a vast system of neurones widely spread over the cerebral cortex and subcortical ganglia. LASHLEY (1950) has convincingly argued that "the activity of literally millions of neurones" is involved in the recall of any memory. His experimental study of the effects of cortical lesions on memory indicates that any particular memory trace or engram has multiple representation in the cortex. Furthermore, LASHLEY concludes that any cortical neurone does not exclusively belong to one engram, but, on the contrary, each neurone and even each synaptic junction would be built into many engrams (cf. Figs. 10, 12). We have already seen that systematic study of the responses of individual neurones in the cortex and in the subcortical nuclei is providing many examples of this multiple operation. Furthermore, physiological, anatomical and biochemical investigations have provided evidence relating to the postulated plasticity of some types of synapses, usage giving growth and disuse regression. These recent discoveries provide a striking corroboration of the original postulates of the synaptic growth theory of learning.

Chapter IV

The Experiencing Self[1]

The Concept of Self

First of all, I am going to talk to you generally about the title for this lecture. And while I do this, I want each one of you to participate with me in an effort to grasp the meaning of what I am trying to say and to apply it to yourself. If you do this with me, we shall have a kind of dialogue in thought, so that my thoughts communicated in language will give you thoughts that parallel mine. You become then not my audience, but my collaborators in this conjoint effort to reach some understanding of what is central to our being, namely the experiencing self. An attempt will be made to see how far we can answer the question: What am I? This is a question which each of us can ask ourselves, and which is quite unashamedly a looking within ourselves—an attitude which is called subjective and introspective.

I am not alone in posing this question and attempting to answer it. For example, SCHRÖDINGER (1951), the physicist who was awarded the Nobel Prize for his wave mechanics, wrote in his book *Science and Humanism:*

> "Who are we? The answer to this question is not only one of the tasks, but the task of science."

This assessment would have been supported by SHERRINGTON, the founder of modern neurophysiology, and many other scientists would be in agreement. I can mention EUGENE WIGNER; CYRIL HINSHELWOOD and MICHAEL POLANYI.

1 This is the text of a lecture delivered to a large student audience on January 12, 1968, at Gustaphus Adolphus College, St. Peter, Minnesota. This Fourth Annual Nobel Conference was on the theme, "The Uniqueness of Man." The original lecture form of the text has been retained, and the text is almost identical with the Conference publication (ECCLES, 1969 a).

I have chosen to talk to you in the field of the philosophy of the person, because I wish to do all I can to restore to mankind the sense of wonder and mystery that arises from the attempt to face up to the reality of our very existence as conscious beings. Too often we have statements that a man is but a clever animal and entirely explicable materially. And again, we are often told that man is nothing but an extremely complex machine and that computers will soon be rivalling him for supremacy as the most complex machine in existence, and that they will have performances outstripping him in all that matters. I want to discredit such dogmatic statements and bring you to realize how tremendous is the mystery of the existence of each one of us.

My approach to conscious experience is, in the first instance, based on my direct experience of my own self-consciousness. I want you to understand that this initial position of myself in regard to my own consciousness must also be adopted by each of you in regard to your own self-consciousness. I am aware that a philosophical position of this kind is often criticized because it is alleged that it gives rise to the exclusive attention of each one of us to our own conscious experiences—an attitude which is called solipsism. However, for a start, I want each of you to face up to the problem discussed in this lecture in this way, and it will soon become apparent that we move from this restricted initial position into the wider field of vision, where we recognize the existence of other conscious selves or persons. That recognition provides, of course, the basis of social life, and it is the denial of solipsism.

You will recognize that, because of memory, each of us links his life together into some kind of continuity of inner experience, which is what we mean when we talk of a self or person. This involves a recognition of unity and identity through all past vicissitudes. Of course, we do not have a continuity of conscious experience. The continuity is broken every time we go to sleep or lose consciousness in some more unpleasant way. But we wake up after each period of unconsciousness, recognizing, because of memory, our continuity with the self of the preceding day, and we continue with our trains of experiences. Is it not a curious experience that, when we wake up in the morning, we slowly come around to recognize that we are in just the same room as when we lost consciousness the night before? Thus we bridge the periods of unconsciousness and identify ourself in the morning with the person who went to sleep the night before. It is the same self that awakes to another stream of consciousness for another waiting day.

The title phrase of this lecture, viz. the experiencing self, relates to the central theme of this conference on the Uniqueness of Man. Man is unique because he alone has come to recognize his existence as a self. For each of us it is central to our experiences as conscious beings, and it is built upon the rich and fascinating tapestry of memories woven from the experiences of a lifetime. Of course, normally we live superficially from occasion to occasion and perhaps mercifully do not often search in depth within ourselves by introspection. But this we certainly do in the crises of life. Much of the greatness of literature is built upon such occasions, as for example, the great soliloquies in Shakespeare's plays, particularly in Hamlet. For example in the "To be or not to be that is the question": there is expressed a stark introspective examination of self.

You will appreciate the concept of self in the use of the words selfish and selfless. At one end of this spectrum we think of examples where an individual is searching assiduously for his own profit and advancement. His relationships with others are simplified into two classes: those that serve his ambition, the pawns in his game of lifemanship; and those who are potentially dangerous (his assumed rivals) or who are obstructions to his planned ascent. The Norwegian sculptor, VIGELAND, has portrayed this struggle of humanity in a great column formed literally by the bodies of people, some trampled down, others climbing up the column of bodies to the summit. At the other extreme there are the selfless who literally do not participate in the game of lifemanship. They are the truly meek and the humble of the Gospel. They must not be confused with the counterfeits so well delineated by C.S. LEWIS. "She is one of those who likes living for others; you can tell the others by the hunted expressions on their faces."

Conscious Experience

So let me now start with this experience that each of us has as a kind of inner illumination, the self-awareness that Professor DOBZHANSKY spoke of in the preceding lecture (1969). I am going to state quite categorically that this conscious experience is all that is given to me in my task of trying to understand myself, and similarly, this is true for each one of you. Further, I am going to state that only because of and through my conscious experience do I come to know of a world of things and events and so to embark on the attempt to understand it; as for example, I do in my work as a scientist. This again is true for each one of you, and in all that you do in your own individual lives.

In developing the significance of this conscious experience, I would like to quote from a recent lecture, "Two kinds of reality," by EUGENE WIGNER (1964). These quotations illustrate how important and urgent is the problem of consciousness to one of the most eminent theoretical physicists in the world today. I quote:

"There are two kinds of reality or existence—the existence of my consciousness and the reality or existence of everything else. This latter reality is not absolute, but only relative. Excepting immediate sensations, the content of my consciousness, everything is a construct: but some constructs are closer, some further, from the direct sensations."

You will see from this quotation that the whole of what we call the material world, that is, the constructs, is regarded by WIGNER as having a second order of reality in contrast to the absolute reality of our conscious experiences, and he develops this theme in the course of his lecture. WIGNER continues:

"As I said, our inability to describe our consciousness adequately, to give a satisfactory picture of it, is the greatest obstacle to our acquiring a rounded picture of the world."

I shall further reinforce this concept of the primacy of our conscious experiences by a quotation from another eminent scientist. Sir CYRIL HINSHELWOOD (1962), in his lecture, "The Vision of Nature," says:

"To deny the reality of the inner world is a flat negation of all that is immediate in existence: to minimize its significance is to depreciate the very purpose of living, and to explain it away as a product of natural selection is a plain fallacy."

The Perceptual World

It is entirely from such perceptual experiences as vision, hearing and touch, for example, that I come to know the external world of things and events, which is a world other than the self of my conscious experience. You may be surprised to hear me say that a special part of this external world is, in fact, my own body, which I actually only come to know because of such senses as vision and touch; and in this same way I come to know of innumerable other human bodies that appear to belong to selves like my own self.

This is, of course, so self-evident that it may appear to you to be a trite statement. Nevertheless, it is of great significance, in that it leads me to believe in the existence of other persons or selves like myself with bodies and conscious experiences. It leads to the rejection of solipsism. From our earliest childhood days, we have learnt to exchange communication with other selves by all kinds of movements or signals. For example, we do this in babyhood by gestures; and, as we become progressively more educated, we use speech and writing; and of course, we learn to exploit still more sophisticated and subtle means of communication, as in the shared joy in aesthetic experience and imagination that even makes words seem too crude.

This takes us into the world of communication by artistic creation and by shared appreciation. But no matter how intimate is our linkage with some dearly-loved person, we still remain separated in a most heart-rending way. We are dependent on some movement such as speech or a gesture that gives to the other a sensory experience. Never does there seem to be a direct communication of one conscious self to the other. At least, I shall say that the direct thought transfer postulated in telepathy appears to be a very inefficient way of communicating between selves. I would not deny the possibility of telepathy, neither do I think it proved. We need further rigorous investigation in this most difficult scientific field.

Thus we come to believe that there is a world of selves, each with the experience of inhabiting a body that is in a material world comprising innumerable bodies of like nature and a tremendous variety of other living forms and an immensity of apparently non-living matter. I would agree with WIGNER (1964) that this material or objective world has the status of a second-order or derivative reality.

I shall now pose two questions: How do I know about this material world in which I must include even my own body? and: How do I know that its existence is not merely in my conscious experience?

When I look at a room with which I am not familiar, such as this lecture hall, I am quite confident that I can assess its dimensions, and that I can verify this judgement from visual experience by an actual exploration if I so choose. Of course, my unaided visual experience could be fooled by the pseudo-rooms built by AMES, which are designed so as to confuse when the observer views them from some particular position. But otherwise, with ordinary rooms, I can ask: How do I recognize their geometrical configuration, when all I have to go on is my vision?

The common explanation is that I have an image on my retina, which is transmitted by nerve impulses in the million-or-so separate nerve fibers in each of my optic nerves, and that the information so travelling to the visual cortex of my brain gives rise to specific patterns of activity, woven in space and time by the activity of the nerve cells. In some mysterious way, this pattern is transmuted into experiences that are projected out into space, and, lo, there is the room visualized by me. This explanation, however, is only partly true. What we can say is that, when there is a specific patterned activation of the nerve cells in my brain, I have a conscious experience that somehow is derived from these specific events. Hence arises the problem: How can this cerebral pattern of activity give me a valid picture of the external world?

Usually, this problem is discussed in relation to visual perception, which is assumed to be an inborn property of the nervous system. On the contrary, the visual world is an interpretation of retinal data, that has been learned through association with information from sense organs, particularly those of muscles, joints, skin and the inner ear, and is the end-product of a long effort of progressive learning by trial and error (cf. DEWEY, 1898).

As a well-trained adult, it is difficult for me to realize that my earliest learning occurred in a cot with the movement of my limbs under visual observation; and thereafter, the field of kinaesthetic and visual education was extended by crawling, walking and still other modes of locomotion, so that my sphere of observation was progressively further enlarged. Thus I would know the dimensions of a visually observed room, because I have crawled, groped, walked and felt all round rooms like it at various stages of my life, and in this way have had my visual impressions built upon these kinaesthetic experiences; and this would be true also for you. I judge distance and space as distance and direction that could be travelled, if I so wish; and so I orientate the world around myself. Thus my three-dimensional perceptual world is essentially a "kinaesthetic world;" it was initially bounded by the cot, but has thereafter been enormously extended in range and subtlety. The learning processes of early childhood are largely unremembered, but I can remember many early efforts to evaluate distance and size, as well as the errors of judgement that I made, when confronted by strange landscapes and seascapes where familiar clues were lacking.

Fortunately, I do not have to rely only on memories from infancy, for in his book *Space and Sight*, von SENDEN (1960) quotes well-documented accounts of adults who were given patterned vision for the first time by the removal of congenital cataracts from their eyes. They reported that their initial visual experiences were meaningless and quite unrelated to the spatial world that had been built up from touch and movement. It took many weeks, and even months, of continual effort to derive from visual experiences a perceptual world that was congruous with their "kinaesthetic world," and in which, as a consequence, they could move with assurance.

A further illustration of the way in which learning can transform the interpretation of visual information is provided by STRATTON'S experiments (1897), in which a system of lenses was placed in front of one of his eyes (the other being covered), so that the image on the retina was inverted with respect to its usual orientation. For several days, the visual world was hopelessly disordered. Since it was inverted, it gave an impression of unreality and was useless for the purpose of apprehending or manipulating objects. But as a result of eight days of continual effort, the visual world could again be sensed by him correctly, and then became a reliable guide for manipulation and movement. There have been several experimental confirmations of Stratton's remarkable findings, and many additional observations, particularly by KOHLER (1951). Subjects with inverted retinal images have even learned to ski, which requires a very accurate correlation of visual with kinaesthetic experiences [2].

These observations and many others of like kind establish that, as a consequence of active or trial-and-error learning, the brain events evoked by sensory information from the retina are interpreted so that they give a valid picture of the external world that is sensed by touch and movement, i.e. my world of visual perception becomes a world in which I can effectively move. It is important to realize (TEUBER, 1966) that we do not learn from a relaxed kaleidoscope of experiences, but rather from what we might call "participation learning." Actually, this perceptual world is much more synthetic than we imagine; for example, it normally remains fixed and stable when the images on the retina are moved in the most diverse ways by naturally occurring body, head or eye movements, but not, for example, when the eye is moved by an applied pressure with the finger on the eyelids either to the side or below the eye. You can try this on yourselves right now, so that you can personally experience the difference between some abnormally-produced movement and natural eye movements.

The kinaesthetic information from all these natural movements, as well as the sensory information from the inner ear receptors for orientation in space, are synthesized with the retinal information. The action of this automatic correction device for visual perception is best appreciated when there are disturbances in the functioning of the inner ear; under such conditions, there is gross movement of the visually-perceived world, which gives rise to the sensation of vertigo.

In our ordinary day to day existence we think that we perceive the visual world objectively, as it really exists, in three dimensions, with all its properties of colour, form and texture. According to this attitude of naive realism we regard our conscious experiences as subjective, private and derivative. However, as I have stated above, the reverse is the case. The primal things are our own conscious experiences. Such perceptual experiences arise because of the coded information that is fed into the

---

2 It may be noted that HARRIS (1965) has presented evidence which he interprets as showing that visual sense completely dominates kinaesthesis in prism distortion experiments. This dogmatic and exclusive interpretation of the recent wealth of experimental data has been criticized, it being postulated that compensation for prism distortion involves an adaptive recalibration of visual and kinaesthetic data (TAUB, 1968). As a neurophysiologist, I readily agree that in the brain there is coordination of all sensory inputs and a reaction that can be relatively dominated by one or the other according to the exigencies of the situation confronting the subject. Moreover all these reactions are tentative, and are continuously subject to revision by dynamic loop operation of all the relevant sensory inputs, there being a loop-time of about 0.1 sec for goal-directed movement (ECCLES, 1970a). There is no justification for the attempt by HARRIS (1965) to extrapolate the artificial situations of prism distortion in the adult to the postulate that in the infant "proprioceptive perception of parts of the body (and therefore of the locations of touched objects) develops with the help of innate visual perception rather than vice versa."

brain by our sense organs, there to produce spatio-temporal patterns of activation of our brain-cells, and which we learn throughout life to interpret in order to give a valid picture of the world in which we live. The criterion of this validity is that we are enabled to act in this world with assurance and success. We tend to overlook the remarkable efforts of learning which enable us to drive a car and negotiate thick traffic at high speed, to judge distances, direction and dimensions instantly, and to act appropriately with skill and finesse, as we do in games, for example.

Therefore, we can now return in a circular manner to the beginning of the story, with an understanding of how the external world can be apprehended by means of our sensory mechanisms, the receptor organs and their pathways to the brain. Moreover, we can communicate to each other with respect to our interpretation of our sensory experiences and discover that to a very large extent we have agreement with one another on these interpretations, which give us what we call the objective world.

The measure of this agreement is perhaps best appreciated by reference to situations where there is disagreement. For example, a considerable number of people differ in their interpretation of colour, and we classify them as colour-blind or colour-defective to varying degrees. Similarly, we have "taste-blindness" (if we may so call it) of many people to a substance, phenylthiocarbamide, which is very bitter to about 75% and tasteless to 25%, and "smell-blindness" of about 18% of males to the extremely poisonous substance, hydrogen cyanide (cf. HUXLEY, 1962).

Again, a subject under the influence of an hallucinogenic drug, such as mescaline or LSD, experiences a wealth of imagery that is not shared by other observers close by. It is readily appreciated that such a discrepancy does not cast doubt on the validity of the external world that is derived from the shared perceptual worlds of these observers; instead, the exceptional experiences that occur under the influence of mescaline or in other disordered cerebral functions are classified as hallucinations. It will be realized that, when observers report one or other of these exceptional features of their perceptual worlds, the situation is customarily handled in a "commonsense" way, so that there is no casting of doubt upon the status of a real external world whose existence and nature is a matter of general agreement so that it can be regarded as being independent of observers.

The example of colour-blindness illustrates in a most pointed manner the amazing dominance of the majority agreement and the commonsense way of repressing the minority of disagreement. This method works out well for the crude levels of perception, using simple criteria of perceptual recognition with nothing more subtle than colour matching; but there are immense divergences in the perceptual experiences of individuals when it comes to such highly sophisticated performances such as occur with philosophical arguments, with aesthetic judgements in music, the plastic arts, and literature, and even with such learned skills as tea and wine-tasting, and the evaluation of design and decor.

And might I add, in a muted tone, these divergences exist amongst scientists in the evaluation and interpretation of experimental data, and in the way in which this data can be used to test scientific hypotheses and so to illuminate our understanding of the natural world. In fact, the tensions and conflicts arising from these differences in interpretation and judgement and belief, give the drive and zest to our creative adventure and performance in both the sciences and the arts.

The Objective-Subjective Dichotomy

It would seem that the status of the external world is assured, for it has a reality that apparently transcends all the imperfections in the perceptual equipments of the observers. In this way there has arisen the contrast between the reality of the external or objective world on the one hand, and, on the other, the subjectivity of our perceptual experiences with all their personal bias and distortion. It is generally believed that the former alone provides a sound basis for scientific investigation. However, this objective-subjective distinction is illusory, being derived from a misinterpretation and a misunderstanding, as has been convincingly argued by SCHRÖDINGER (1958); for example, he says:

> "Without being aware of it and without being rigorously systematic about it, we exclude the Subject of Cognizance from the domain of nature that we endeavour to understand. We step with our own person back into the part of an onlooker who does not belong to the world, which by this very procedure becomes an objective world. This situation is the same for every mind and its world, in spite of the unfathomable abundance of 'cross-references' between them. The world is given to me only once, not one existing and one perceived. Subject and object are only one. The barrier between them cannot be said to have broken down as a result of recent experience in the physical sciences, for this barrier does not exist."

The illusory nature of the objective-subjective dichotomy of experience is further illustrated by what might be called a spectrum of perceptual experiences. (A) The vision of an object is confirmed by touching it, so giving it form, and in this same manner it can be sensed by and reported upon by other observers, the perception of the object thus achieving a public status. (B) Pin prick of a finger can be witnessed by an observer as well as by the subject, but the pain is private to the subject; however, each observer can perform a similar experiment on himself and report his observation of pain, which in this way is shared and so achieves a public status. (C) The dull pain or ache of visceral origin cannot be readily duplicated in another observer, yet clinical investigators have provided a wealth of evidence on the pains characteristic of the visceral diseases, and even of referred pains, so that reports of visceral pain achieve indirectly a kind of public status. Similar considerations apply to such sensations as thirst or hunger. (D) Unlike the preceding three examples, mental pain or anguish is not a consequence of stimulation of receptor organs; yet again a kind of public status can be given to such purely private experiences, for there is a measure of agreement in the reports of subjects so afflicted. Similar considerations apply to other emotional experiences: anger, joy, delight in beauty, awe, fear. (E) The experiences of dreams or of memories are even more uniquely private, belonging as they do still more exclusively to the realm of inner experience; yet again a kind of public status is established by the wealth of communication that there is between observers.

It can be claimed that all transitions exist between any two successive examples of this spectrum, which conforms very well with the postulate that every one of the various experiences is associated with specific patterns of neuronal activity in the brain. Apparently such specific patterns can sometimes be evoked by electrical stimulation of the brains of epileptic patients. PENFIELD and JASPER (1954) and PENFIELD (1968) have given fascinating accounts of the way in which electrical stimulation of the temporal lobe of the cerebrum will evoke vivid and detailed memories of long-forgotten incidents.

The conclusion is that every observation of the so-called objective world depends in the first instance on an experience which is just as private as the so-called subjective experiences. The public status of an observation is given by symbolic communication between observers, in particular through the medium of language. By means of this same method of communication, our inner or subjective experiences can likewise achieve a public status.

When I re-examine the nature of my sensory perceptions, it is evident that these give me the so-called facts of immediate experience and that the so-called external world or "objective world" is a derivative or representation of certain types of this private and direct experience. But this "representative theory of perception" (BELOFF, 1962) must not be confused with idealist monism, for the implication is that my perceptual world is my symbolic picture of the "objective world" and thus resembles a map. This map or symbolic picture is essential so that I can act appropriately within this "objective world;" and, as we have seen, it is synthesized from sensory data so as to be effective for this very purpose. It is built upon spatial relations, but is also given symbolic information in terms of secondary qualities. For example, colours, sounds, smells, heat, and cold, as such, belong only to the perceptual world.

Some speculative glimpses of neuronal operation can be achieved by taking into account the fact that many almost synchronous synaptic excitatory bombardments are essential for causing any cell to generate an impulse and thus to contribute to the further spread of neuronal activity. For an effective spread of activity, each neurone must receive synaptic activation probably from hundreds of neurones and itself transmit to hundreds of others. Thus there arises the concept of a wave-front, as illustrated in Figs. 10 and 12, comprising a kind of multi-lane traffic in hundreds of neuronal channels, so that the wave-front would traverse 100,000 neurones in one second, weaving a kind of pattern in space and time in a way that SHERRINGTON (1940), with his poetic insight, has likened to the operations of an "enchanted loom." Furthermore, when, by means of a microelectrode, single cerebral neurones are being investigated, it is often found that one can be activated from several different sensory inputs (Fig. 11).

It may help in this way to think of the nervous system as an enormously complicated telephone exchange, constructed from 10,000 million unitary components or nerve cells, but, of course, operating in a way that is very different from a telephone exchange because of the necessity for summation that gives our brain its great ability in the correlation of the data received through the enormous number of channels leading from any one sense organ. For example, there are about one million separate nerve fibers from each eye. Also there is the correlation of the information that comes in from different sense organs, such as we have when we are using our eyes and our sense of touch to guide and control our movement, or when we correlate something that we see with something that we hear.

Within the last decade there have been great advances in studying the cerebral cortex at very high magnifications in electronmicroscopy (Figs. 3B, 4A, B, 18, 32C), and also in employing electrical recording from individual nerve cells in order to study the mode of action of the synaptic contacts which they make with each other (Figs. 5, 13, 14). However, as a result of this intense scientific study, we are still only at the first stage of understanding the events concerned in perceptual awareness, and have virtually not approached at all the more complex problems of perceptual recognition and judgement.

There is much neurophysiological evidence to show that a conscious experience arises only when there is some specific cerebral activity. For every experience it is believed that a specific spatio-temporal pattern as in Fig. 12 is woven by the "enchanted loom" of nerve cell activation in the brain (cf. FESSARD, 1961). With sensory perception the sequence of events is that some stimulus to a sense organ causes the repetitive discharge of impulses along sensory nerve fibers to the brain (cf. Figs. 6, 31A), which, after various synaptic relays therein (Figs. 7, 8, 9), eventually evoke specific spatio-temporal patterns of impulses (Figs. 11, 33A) in the neuronal network of the cerebral cortex (Figs. 10, 12). The transmission from sense organ to cerebral cortex utilizes a coded pattern of nerve impulses that may be likened to a Morse Code with dots only in various temporal sequences. Certainly, this coded transmission is quite unlike the original stimulus to that sense organ, and the spatio-temporal pattern of neuronal activity that is evoked in the cerebral cortex by this weaving in the "enchanted loom" (cf. Fig. 12) is again quite different. Yet, as a consequence of these cerebral patterns of activity in my brain, I experience sensations which I have learned to project to somewhere outside the cortex; it may be to the surface of my body or even within it, but most commonly, as with sight, hearing or smell, to the outside world.

It cannot be too strongly emphasized that this investigation into the neuronal mechanism of the cerebral cortex is still at a primitive stage, and hence gives but some dim and shadowy picture of the amazing intricacy of pattern woven in space and time by the sequential activation of neurones in multi-lane traffic over the 10,000 million components in the cortical slab of cells. It has been surmised that many millions of cells take part in the cortical response leading to the simplest conscious experience. We can further speculate that the human cerebral cortex surpasses that of all other animals in its potentiality to develop subtle and complex neuronal patterns of the utmost variety, for from this would stem the richness of human performance as compared with that of even the most intelligent animal.

This direct relationship of brain activity to our perceptual experiences was first clearly seen by the French scientist and philosopher, DESCARTES,[3] in the seventeenth century. Though he was wrong in all details of his explanations of brain action, he was right in essentials. It is immaterial whether brain events are caused by local stimulation of the cerebral cortex or some part of the sensory nervous pathway, as in the phantom

3 As well as being a philosopher, DESCARTES was a doctor. One of his discoveries related to the nature of the phantom limb by carrying out an experiment that was quite remarkable in the seventeenth century. Up till that stage, people had thought that the phantom limb phenomena arose because the patients with amputated limbs wanted to imagine that their limbs were still intact. Without telling her what he was doing, DESCARTES had the necessary amputation performed on the very badly gangrenous arm of a girl, after tying a tourniquet to render the limb insensitive. Dressings and bandages were arranged so that she might think that she still had her forearm and hand. She regained limb sensation and felt that the hand still there, even describing how each individual finger felt.

DESCARTES attributed this to the stimulation of nerve endings in the amputated stump, these stimuli reaching the brain, which was instrumental in interpreting the nerve stimuli as if they had come, as normally, from the limb; hence the illusion of the phantom limb. This explanation is essentially that accepted at the present day. (DESCARTES, R., 1644, The Principles of Philosophy, Part 4, Principle 196, Translation, 1931.)

limb, or whether it is, as is usual, generated by impulses discharged by sense organs. However, as reported by PENFIELD (1966), electrical stimuli applied to the sensory zones of the cerebral cortex usually evoke only chaotic sensations: tingling or numbness in the skin zones; lights and colours in the visual zone; noises in the auditory zone. Such chaotic responses are to be expected, since electrical stimulation of the cortex must directly excite the tens of thousands of neurones regardless of their functional relationships, and so initiate a widely spreading amorphous field of neuronal activation, very different from the fine and specific patterns that must be set up by the input to the cortex from the sense organs. A familiar chaotic sensation, involving elements of touch, heat, cold and pain, arises for a similar reason when a sensory nerve is directly excited, as when the ulnar nerve in the elbow (the "funny bone") is stimulated by a sudden blow.

As will be described in the next chapter, the time required for the elaboration of the neuronal substrate of a very simple conscious experience may be at least one fifth of a second (LIBET, 1966; CRAWFORD, 1947). Transmission from one nerve cell to another takes no longer than one-thousandth of a second; hence, there could be a serial relay of as many as two hundred synaptic linkages between nerve cells before a conscious experience is aroused. Many thousands of nerve cells would be initially activated, and each nerve cell by synaptic relay would in turn activate many nerve cells so that millions would be activated in a fifth of a second.

The immensity of this patterned spread throughout the neuronal pathways formed by millions of nerve cells in the brain can be best imagined by thinking of the patterns of Medieval tapestries or of Oriental carpets. But as SHERRINGTON (1940) says with his rare insight, the loom weaves "a dissolving pattern, always a meaningful pattern, though never an abiding one; a shifting harmony of subpatterns." This tremendous complication of neuronal activity in my brain is required before a sensory input is perceived by me even in the rawest form; and responses involving comparison, value, judgement, correlations with remembered experiences, aesthetic evaluations, undoubtedly take much longer, with the consequence that there must be quite fantastic complexities of neuronal operation in the spatio-temporal patterns woven in the cerebral cortex.

There is now evidence that at any one time we are only conscious of an extremely small fraction of the immense sensory input that is pouring into our brains (MORUZZI, 1966a). In fact, by far the greater part of the activity in the brain, and even in the cerebral cortex, does not reach consciousness at all. However, we have the ability to direct our attention apparently at will to one or another element in the input from our sense organs.

To continue with our subjective impressions, I now refer to dreams, which, like memories, must involve retracking through some specific spatio-temporal patterns of neuronal activity which were the basis of various perceptions in the past. At intervals of two or three hours every night we have a dream cycle (KLEITMAN, 1961, 1963). This can be shown by movements of a subject's eyes, which are revealed by electrical recording from the eyelids. Thus we can be observing a subject who is asleep, and we notice these movements of his eyes. If we then immediately awaken him, he will report the dream. If, instead, he is allowed to sleep on and is awakened some ten minutes later, he will say that he has not dreamed. Evidently, he has forgotten the dream. At the time of the dream there is special brain activity, as revealed by its electrical activity that is recorded as an electroencephalogram.

We can conclude then that, when there are some organized patterns of nerve cell activity evolving in some sequential manner, there may be a conscious experience, be this a dream, or, if we are awake, a daydream, a memory, a perceptual experience or a thought.

Another problem relates to attention, since much organized cortical activity can be carried on at a subconscious level. How does it come about that our attention is diverted now to this patterned activity which then gives its unique conscious experience, now to that? Is the magnitude of the cortical activation of significance in thus achieving conscious attention? These fundamental problems remain as yet beyond more than the vaguest formulation of fragmentary hypotheses.

Our speculation has been extended to cover in principle the simplest aspects of imagination, imagery or the re-experiencing of images. In passing beyond this stage we may firstly consider a peculiar tendency to association of imagery, so that the experience of one image is evocative of other images, and these of still more and so on. When these images are of beauty and subtlety, blending in harmony and capable of being expressed in some language, verbal, musical or pictorial, so that transcendent experiences can be evoked in others, we have artistic creation of a simple or lyrical kind. Alternatively, entrancing displays of imagery that are reputed to be of great beauty and clarity can be experienced by ordinary people under the influence of hallucinogenic drugs such as mescaline or LSD. In parenthesis it should be noted that there are very few transmutations into literature or art of the transcendent aesthetic experiences alleged to be enjoyed by drug addicts. One would suspect that in these conditions there would be an especial tendency for the formation of ever more complex and effectively interlocked patterns of neuronal activity involving large fractions of the cortical population of neurones. This would account for the withdrawal of the subject from ordinary activities during these absorbing experiences. Not unrelated to these states are the various psychoses where the inner experiences of the patients also cause them to be withdrawn.

Cerebral Events and Conscious Experience

I now return to the key problem in perception which can be expressed in the question: How can some specific spatiotemporal pattern of neuronal activity in the cerebral cortex evoke a particular sensory experience? We can dimly perceive a relationship between brain states and consciousness when we consider the neuronal activity of the cortex in states of unconsciousness, that is, when stimulation of sense organs fails to evoke a sensory experience. The electroencephalogram reveals that in such states there may be either a very low level of neuronal activity, as in coma, concussion, anaesthesia, and deep sleep, or a very high level of stereotyped and driven activity, as in convulsions. On the contrary, the electrical activity of the awake brain indicates that a large proportion of the neurones is occupied in an intense dynamic activity of great variety (cf. FESSARD, 1961). Under such conditions it has been postulated that at any instant a considerable proportion of the neurones would be passing through levels of excitation at which the discharge of an impulse would exhibit uncertainity, such neurones being "critically poised" with respect to the generation of impulses (cf. Chapter VIII; ECCLES, 1953, Chapter 8). Diagrams of activated neuronal networks as in Figs. 10 and 12 can form the basis of imaginative constructs of spatio-temporal patterns that would develop if such inputs were superimposed on high levels of background activity. It has further been postulated that consciousness is dependent on the existence of a sufficient number of such critically poised neurones, and, consequently, only in such conditions are willing and perceiving possible (cf. Chapter VIII). However, it is not necessary for the whole cortex to be in this special dynamic state. There is clinical evidence that excision of a large part of the cerebral cortex does not interrupt consciousness; and in convulsions, unconsciousness does not supervene until the convulsive activity has invaded a large part of the cortex. Furthermore, I would suggest that the transcendent performance of the central nervous system is a consequence of its amazing complexity, not only structural, but also dynamic, which is of a much higher order than any other organized system in the universe.

On the basis of this concept we can face up anew to the extraordinary problems inherent in a strong dualism. Interaction of brain and conscious mind, brain receiving from conscious mind in a willed action and in turn transmitting to mind in a conscious experience. But let us be quite clear that for each of us the primary reality is our consciousness—everything else is derivative and has a second order reality. We have tremendous intellectual tasks in our efforts to understand baffling problems that lie right at the center of our being; but as EUGENE WIGNER (1964) asks: "Have we any right to expect a solution to such fundamental problems when the efforts made have been trivial relative to the extreme nature of the problem?"

The Principles of Emergence

Evidently the unimaginable organized complexity of the cerebrum has caused the emergence of properties which are of a different kind from anything that has been as yet related to matter with its properties as defined in physics and chemistry.

In order to follow up this challenging insight it is essential to trace the hierarchial sequences from our concepts of the fundamental properties of matter that is the subject matter of physics and chemistry. Firstly we shall consider a machine such as a watch. Any machine is defined by its operational principles, which include its structure and the manner and purpose of its operation. POLANYI (1966, p. 40) raises the questions:

> "But how can a machine which, as an inanimate body, obeys the laws of physics and chemistry fail to be determined by these laws? How can it follow both the laws of nature and its own operational principles as a machine? How does the shaping of inanimate matter in a machine make it capable of success or failure? The answer lies in the word: shaping. Natural laws may mould inanimate matter into distinctive shapes, such as the spheres of the sun and the moon, and into such patterns as that of the solar system. Other shapes can be imposed on matter artificially, and yet without infringing the laws of nature. The operational principles of machines are embodied in matter by such artificial shaping. These principles may be said *to govern the boundary conditions of an inanimate system*—a set of conditions that is explicitly left undetermined by the laws of nature. Engineering provides a determination of such boundary conditions. And this is how an inanimate system can be subject to a dual control on two levels: the operations of the upper level are artificially embodied in the boundaries of the lower level which is relied on to obey the laws of inanimate nature, i.e., physics and chemistry.
>
> We may call the control exercised by the organizational principle of a higher level on the particulars forming its lower level *the principle of marginal control.*"

POLANYI (1967a) develops these concepts further in relation to books and other means of communication.

> "Nothing is said about the content of a book by its physical-chemical topography. All objects conveying information are irreducible to the terms of physics and chemistry."

But of particular interest is his statement of the general principle of emergence in respect to life (POLANYI, 1966, p. 44; cf. POLANYI, 1968b):

> "But the hierarchic structure of the higher forms of life necessitates the assumption of further processes of emergence. If each higher level is to control the boundary conditions left open by the operations of the next lower level, this implies that these boundary conditions are in fact left open by the operations going on at the lower level. In other words, no level can gain control over its own boundary conditions and hence cannot bring into

Nevertheless, the truth of the two kinds of reality seems irrefutable. Will it ever be possible to resolve and understand this desperately unsatisfactory conflict between known phenomena and our expectations? We do not know. However, if it will be possible to 'understand' the awakening of the consciousness at birth, and its extinction at death, it will be possible through a study of these phenomena on a broad scale. It would be contrary to all our past experience with science, if we had understood, with as perfunctory an effort as we have made so far, the phenomena most deeply affecting the realities of the first kind."

Certainly one of the poignant problems confronting each man in his life is his attempt to become reconciled with his inevitable end in death. This, of course, is relatable to his evolutionary origin. He dies as do other animals, but the inevitability of death afflicts man alone because man in his development gained self-consciousness (DOBZHANSKY, 1967, 1969). The recognition of this terrible problem of death-awareness has been responsible for the myths and religions that have been developed very largely to give man assurance in facing up to the ultimate end of his brief life. Now that man has so far forsaken these myths and religions he feels frightened and alone. However, there is so much unknown in our origin as individual conscious beings, and this origin transcends our evolutionary origin; hence we must not infer that life has no meaning beyond the drama played out on this earth. As I pointed out in my lecture at this conference last year (ECCLES, 1967) and also in my Eddington Lecture (ECCLES, 1965a) the evolutionary origin of our bodies and their building by the unique instructions provided by DNA inheritance is at best but a partial explanation, and certainly not a sufficient explanation of our existence as conscious beings.

So I come to the conclusion that each of us as an experiencing being exemplifies the essential uniqueness of man. Man's attempt to understand the world is a measure of his uniqueness, but he has been misled by leaving himself, as an experiencing being, out of the totality of the experiences for which he develops scientific and philosophical explanations. I should like to close by indicating that Socrates had a similar idea of the self and the body. The quotation is from the *Phaedo*.

"We shall try our best to do as you say," said Crito.

"But how shall we bury you?"

"Anyway you like," replied Socrates, "that is, if you can catch me and I don't slip through your fingers." He laughed gently as he spoke, and turning to us went on: "I cannot persuade Crito that I am this Socrates here who is talking to you...; he thinks that I am the one whom he will see presently lying dead.... You must give an assurance to Crito for me... that when I am dead I shall not stay but depart and be gone. That will help Crito to bear more easily... when he sees my body being burned or buried, as if something dreadful were happening to me.... No, you must keep up your spirits and say that it is only my body that you are burying; and you can bury it as you please...."

Chapter V

The Brain and the Unity of Conscious Experience[1]

The Reality of Conscious Experience

I believe that this problem that I am talking to you about today would have been one of particular interest to Sir ARTHUR EDDINGTON. Repeatedly in his books he made reference to the problem of consciousness in relation to the physical world that he spoke about with such imagination and such understanding. I can instance his attitude to conscious experience by two brief quotations from his Swarthmore Lecture (1929), *Science and the Unseen World:*

> "In comparing the certainty of things spiritual and things temporal, let us not forget this—Mind is the first and most direct thing in our experience; all else is remote inference.
>
> Picture first consciousness as a bundle of sense-impressions and nothing more.... But picture again consciousness, not this time as a bundle of sense-impressions, but as we intimately know it, responsible, aspiring, yearning, doubting, originating in itself such impulses as those which urge the scientist on his quest for truth."

And in his great book *The Philosophy of Physical Science* (1939, p. 195) he states:

> "The only subject presented to me for study is the content of my consciousness. According to the usual description, this is a heterogeneous collection of sensations, emotions, conceptions, memories, etc. The raw materials of knowledge and the manufactured products of intellectual activity exist side by side in this collection."

And his further statement (p. 206) that "the unity of consciousness is manifested *because* there are parts for it to unite" is of particular relevance to my theme today.

But EDDINGTON was not alone amongst the great physicists of this century in recognizing the importance and urgency of the problem of consciousness. For example, I can instance SCHRÖDINGER'S contributions in his monographs, *Science and Humanism* (1951) and *Mind and Matter* (1958), and EUGENE WIGNER'S (1964) lecture, *Two Kinds of Reality.*

---

1 This Chapter is in large part a reprint of the 19th Eddington Memorial Lecture (ECCLES, 1965a) that I delivered at Cambridge University on October 15th, 1965.

In contrast, until recently it has been fashionable for philosophers and psychologists to discredit or even to deride all problems purporting to derive from the concept of mind or of consciousness (cf. RYLE, 1949). However, the recent reaction to this obscurantism can be illustrated by such books as *The Existence of Mind* (BELOFF, 1962) and *On Having a Mind* (KNEALE, 1962). For example, BELOFF begins his book by stating:

"The thesis of this book, if it can be stated in two words, is that Mind exists, or to be more explicit, that minds, mental entities and mental phenomena, exist as ultimate constituents of the world in which we live;" and also: "Those who take seriously the existence of Mind are often taunted with being worried by a 'ghost in the machine;' I suggest it is high time we refused to let our critical faculties be paralysed any longer by this pert gibe."

This counter-attack is very encouraging to neurophysiologists and neurologists, for many of us, despite the philosophic criticisms, have continued to wrestle with the problem of brain and mind, and have come to regard it as the most difficult and fundamental problem concerning man. Nevertheless, despite this clarification I feel that there is still confusion in the use of such words as mind, mental, mentality, which in some extremely primitive form are even postulated as being a property of inorganic matter! Hence I have refrained from using them, and employ instead either "conscious experience" or "consciousness."

There is a unity of the self through all the diverse conscious experiences of one's life, each of which is assimilated to the self; and this even occurs in dreams, as I think you will readily agree. For always in a dream we find ourselves as the agent central to the whole play of imagery; and this likewise occurs with hallucinations and the fantasies of waking life that we may call daydreams. I want you to recognize that each of you can look back in memory through, as it were, the thread of the long years of accumulated experiences that make up your life, so that eventually you come to your earliest memories, where you have the amazing experience in retrospect of waking up in life in the very limited environment of a young child. Each of you has a personal identity from these earliest times, which is built up from remembered experiences.

I will further suggest that in literature we have not just a description of the behaviours of people going through motions in some determined and stereotyped manner and observed always from the outside. But instead—central to literature—there are descriptions of inner experiences with thoughts and motives and the emotional feelings of the characters that the author, as it were, brings to life in this way. You can yourself recollect all the range of emotional feelings of love and friendship and hate and antipathy, as well as your experiences of fear and terror and of delight in the beautiful. All of these contribute to the richness of your direct inner experiences.

This richness of our experiences is enormously developed, when we fuse, as it were, our immediate perceptual experiences with an imaginative range of inner experiences. This occurs particularly in aesthetic experiences. The artist attempts to express not some exact rendering of what he sees, but his vision, which has a creative enrichment given by his imagination. In great art this artistic creation has for the artist some compelling necessity. Unfortunately, much imitative work masquerades as art, though in itself it is only an artificial contrivance. Superficially, there may be, of course, resemblance between this imitative contrivance and true artistic creation. For this reason, in attempting aesthetic evaluation and understanding, we often need art critics and historians to guide and inform us—but not, let me add, to compel us.

SHERRINGTON (1940) in his Gifford Lectures *(Man on his Nature)* has written most movingly on the self:

> "This 'I', this self, which can so vividly propose to 'do', what attributes as regards 'doing' does it appear to itself to have? It counts itself as a 'cause'. Do we not each think of our 'I' as a 'cause' within our body? 'Within' inasmuch as it is at the core of the spatial world, which our perception seems to look out at from our body. The body seems a zone immediately about that central core. This 'I' belongs more immediately to our awareness than does even the spatial world about us, for it is directly experienced. It *is* the 'self'."

Perceptual Experience

In contrast to this inner experience, I have experiences or perceptions that are derived from activation of my sensory receptors. It is solely from such perceptual experiences that I derive the concept of an external world of things and events, which is a world other than the world of my inner experience. Furthermore, it is part of my interpretation of my perceptual experience that my 'self' is associated with a body that is in the objective world; and I find innumerable other bodies that appear to be of like nature. I can exchange communications with them by bodily movements that give rise to perceptual changes in the observer, for example by gestures or, at a more sophisticated level, by speech that is heard or by writing that is read, and thus discover by reciprocal communication that they, too, have conscious experiences resembling mine. Solipsism becomes for me no longer a tenable belief. EDDINGTON (1939, p. 198) makes a valuable statement on this theme:

> "Thus recognition of sensations other than our own, though not required until a rather later stage of the discussion, is essential to the derivation of an *external* physical universe. Our direct awareness of certain aural and visual sensations (words heard and read) is postulated to be an indirect knowledge of quite different sensations (described by the words heard and read) occurring elsewhere than in our own consciousness. Solipsism would deny this; and it is by accepting this postulate that physics declares itself anti-solipsistic."

As a result of this intensive perceptual training over the years of our lives and of its concentration in the methods of scientific investigations, we have come to learn about sense organs and brains. Gradually, by scientific experiments, the primitive concepts of their modes of operation in perception have become better understood, both the mode of operation of the sense organs as highly specific detectors of physical or chemical stimuli and the way in which information is communicated as signals (nerve impulses) from them to the cerebral cortex; but you will appreciate that this understanding has been developed as a consequence of highly complex and specific intellectual processes—thinking, observing, assessing, correlating, criticizing, reasoning, imagining.

In this city of KEITH LUCAS, ADRIAN, RUSHTON, HODGKIN, HUXLEY and KEYNES, it should not be necessary for me to tell you anything about the nerve impulse, because it is in Cambridge pre-eminently that the fundamental work has been done on this basic mode of communication in the nervous system. However, I must now give you a brief glimpse into the general structural and functional characteristics of the central nervous system.

The human cerebral cortex is a sheet of about 2000 sq. cm in area and about 3 mm in thickness. It is formed by a very dense packing of nerve cells of many varieties and sizes, in all about ten thousand million, and it is often likened to a vast telephone exchange. Within the last few years there have been enormous advances in studying the cerebral cortex by electron microscopy and in employing both intracellular and extracellular recording from the pyramidal cells in order to study the way in which the nerve cells communicate with each other by means of the synaptic contacts (cf. Chapter II). An impulse discharged from one cell causes a momentary activation of the many excitatory or inhibitory synapses that each cell forms with other cells, often many hundreds. Some speculative glimpse of neuronal operation can be achieved by realizing that many almost synchronous excitatory synaptic bombardments are essential for causing any cell to generate an impulse and itself thus to contribute to the further spread of neuronal activity (Fig. 5). For an effective spread of activity each neurone must receive synaptic activation probably from hundreds of neurones and itself transmit to hundreds of others. One is thus introduced to the concept of a wavefront (cf. Fig. 10) comprising a kind of multi-lane traffic in hundreds of neuronal channels, so that the wavefront would sweep over at least 100,000 neurones on one second. Furthermore, there is a great deal of evidence to show that a particular neurone may participate in the patterns of activity developing from many different inputs (cf. Fig. 11). The mode of operation of pathways concerned in perception has been described in Chapters II and IV, with the transmission of information from receptor organs in the coded form of impulse traffic. However, as succinctly expressed by RUSSELL BRAIN (1951), 'the only necessary condition of the observer's seeing colours, hearing sounds, and experiencing his own body is that the appropriate physiological events shall occur in the appropriate areas of his brain.'

The Threshold of Conscious Experience

It has long been known that sensations can be evoked by electrical stimulation of the brain of conscious subjects, and a most thorough investigation has been made by PENFIELD and his associates (PENFIELD and JASPER, 1954). Usually these sensations are disordered experiences or paraesthesia; light or colours from the visual area, tingling and numbness from the somaesthetic area (the area concerned in general body sensations); noises from the auditory area. Recently LIBET and his colleagues (1966) have utilized these responses of the somaesthetic area in an attempt to discover the nature of the neuronal activity that leads to a conscious experience. They applied very weak trains of brief electric pulses, usually at frequencies of 30—60 a second, to the exposed cerebral cortex of conscious subjects who had generously volunteered time during some therapeutic brain surgery. The object of the experiment was to determine the stimulus that just sufficed to cause them to report that they had a conscious experience, which was of course of somaesthetic character, usually abnormal, but in about one third it was normal—for example, a sensation of pressure or touch or movement, and even heat or cold. It was of great interest that, as the number of stimuli in the train increased, there was a large reduction in the strength required to produce a conscious experience, and that at least half a second of repetitive stimulation was required for the weakest stimulus. As shown in Fig. 21 this effectiveness of duration was much the same for 60/sec stimulation as for 30/sec. Continuation of the weakest stimulus beyond the time for production of a sensation did not increase the sensation, but merely prolonged it at the threshold of feeling.

There would be general agreement that each electrical stimulus of the train would be exciting the discharge of impulses from nerve cells and that the effect of duration in lowering threshold strength indicates that there must be an elaboration of the spatio-temporal patterns of impulse discharges before a conscious experience arises. Furthermore, it is suggested that, with all conditions of threshold stimulation, there is a delay of at least half a second before the onset of the experienced sensation. Fig. 22 shows that even the first stimulus of the train evoked an electrical

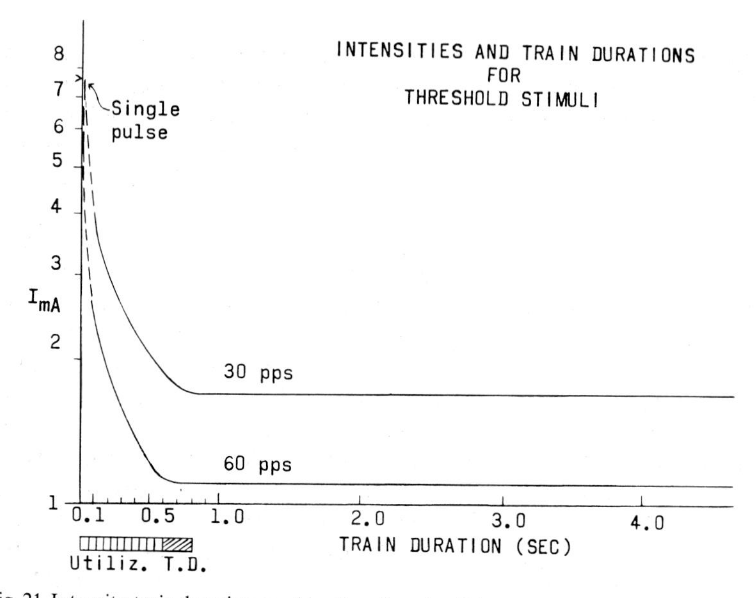

Fig. 21. Intensity-train duration combinations for stimuli (to postcentral gyrus) just adequate to elicit a threshold conscious experience of somatic sensation. Curves are presented for two different pulse-repetition frequencies, employing rectangular pulses of 0.5 msec duration and at the intensities indicated in mA. If the duration of the train is shortened to less than the Utiliz. T.D., a stronger stimulus is required to evoke a threshold response (LIBET, 1966)

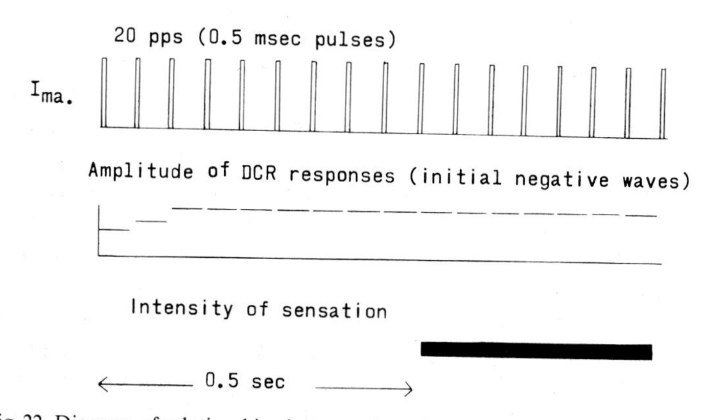

Fig. 22. Diagram of relationships between the train of 0.5 msec pulses at liminal intensity applied to postcentral gyrus, and the amplitudes of the direct cortical responses (DCR) recorded nearby. The third line indicates that no conscious sensory experience is elicited until approximately the initial 0.5 sec of events has elapsed and that the just-detectable sensation appearing after that period remains at the same subjective intensity while the stimulus train continues (LIBET, 1966)

response of the cortex and, after the first three stimuli there was no further increase in this response, though the conscious experience did not arise until after the tenth stimulus. Evidently there is opportunity for a great elaboration of neuronal activity in complex spatio-temporal patterns during the 'incubation period' of a conscious experience at threshold level.

These same temporal characteristics are exhibited with stimulation applied to subcortical white matter or to the thalamus, so it can be assumed that there is an essential factor of elaboration of activated neuronal patterns before a sensation is experienced. This inference conforms with the finding of JASPER (1966) that the initial electrical responses produced by afferent volleys to the cortex are not related to conscious experience. For example, these initial responses are unaltered in relatively deep anaesthesia, but following the initial response there are small after-waves for a second or more and these are very sensitive to the depth of anaesthesia, and are in fact correlatable with the experiencing of sensation.

With the visual system also there is evidence (CRAWFORD, 1947) that at least 0.2 sec of cortical activity is required before a just threshold flash of light can be detected. This elaboration time for a conscious experience may be as long as 1 sec; so a sensory input may evoke quick motor reactions by operation at subconscious levels before it is actually experienced. It is important to recognize that measurements of reaction time cannot be used as a measure of the time required for elaboration of a conscious experience.

The Neuronal Activity Concerned in Conscious Experience

The time of at least 1/5th of a second for the elaboration of the neuronal substrate of a conscious experience is very long indeed. The time for transmission from one nerve cell to another is no longer than 1/1000th of a second; hence there could be a serial relay of as many as 200 synaptic linkages between nerve cells before a conscious experience is aroused. Many thousands of nerve cells would be initially activated, and each nerve cell by synaptic relay would in turn activate many nerve cells. The immensity of this patterned spread throughout the neuronal pathways of the brain is beyond all imagining. This tremendous complication of neuronal activity in my brain is required before a sensory input is perceived by me even in the rawest form; and responses involving comparison, value, judgement, correlations with remembered experiences, aesthetic evaluations undoubtedly take much longer, with the consequence that there must be quite fantastic complexities of neuronal operation in the spatio-temporal patterns woven in the 'enchanted loom,' to use the phrase of SHERRINGTON.

MOUNTCASTLE (1966a) emphasizes that we still know very little about the perceptive mechanism, but that we have made considerable progress in respect to both its neural substratum and its behaviour output. Moreover, as illustrated by the experiments of LIBET and his group (1966), there is a steady closure between these two fields. However, MOUNTCASTLE believes:

... that studies on the cerebral cortex have revealed only the general framework upon which much more complex activities must depend .... At some stage in the long chain of events from stimulus to introspective evaluation, the signs of place and quality must be coded in some more efficient way. The mechanism eludes us; there are not even—to my knowledge—reasonable models for experimental test. I cite this as one of the emergent properties of populations of neurons, one which I see no way of understanding as a simply additive property of the function of single cells, so far as we presently know them.

MOUNTCASTLE (1966b) further goes on to state that:

... The truth is that so far physiological studies have only added to what we know of the anatomical substratum for function—and this is what I mean by the term *static properties.* The tide of experiment moves towards an elucidation of the time-dependent, dynamic aspects of cortical function. One may suppose, and indeed some observations already indicate, that the sequential forming and reforming of new and highly complex patterns of activity, occupying both the cortical cells activated initially by a sensory stimulus and others independent of it, results in functional patterns far more complex than those predicted by the columnar organization alone, and it is this aspect of neural activity to which I believe we must look for the neural correlates of the perceptive process.

An attempt to show schematically such functional patterns of neurons can be seen in Figs. 10 and 12.

In support of his above statement, MOUNTCASTLE cites the remarkable finding of PENFIELD and others that electrical stimulation of a cortical sensory area not only produced the paraesthesiae that are characteristic of that area, but also silences its normal function. The artificially produced patterns of neuronal activity must interfere with the development of the complex "meaningful" patterns that underlie a conscious experience. He concludes that (1966b)

"integrative action, neural discrimination, perception, and perhaps conscious action itself may be regarded as emergent properties of large populations of neurons, properties to be revealed only by continuing experiment."

MORUZZI (1966b) presents an impressive amount of data in support of the postulate that only an extremely small fraction of all the sensory input is actually experienced—for example, during sleep there is a continuing overall activity of cortical nerve cells, and EVARTS (1964) finds that there may even be increased activity of some cortical neurones. Again it will be recognized that a very small fraction of the immense volume of visual information is utilized in visual perception from moment to moment. MORUZZI suggests that only an extremely small proportion of all the patterned neuronal activity that is going on in a brain at any one time gives rise to conscious experience, though, within limits, we may apparently direct our attention to other neuronal patterns which, as a consequence, are then consciously experienced. Likewise, only an extremely small fraction of the sensory input into the brain can be recalled in the process of memory, even for the few minutes of short term memory. In summary, MORUZZI (1966 b) states that:

> "All these considerations lead to the conclusion that the neural processes underlying learning and forgetting, storage and retrieval of memory traces are quantitatively small with respect to the background activity of the cerebrum, although the highest achievements of mankind, from artistic creation to scientific discovery, are dependent upon them."

The great sub-cortical nuclei, and particularly the reticular activating system, have been shown by MAGOUN and MORUZZI and by BREMER to arouse or energize the cortex by a continual barrage of impulse discharge, so maintaining levels of cortical awareness. However, though this background or supporting activation is essential to the maintenance of consciousness, it must be distinguished from the process of attention which operates in some more selective manner than can be provided by these non-specific systems. In agreement with MOUNTCASTLE, I would regard all the fine grain of conscious perception that we experience during attention as essentially arising in the cortex.

Unity of Conscious Experience and the Cerebral Commissures

There are remarkable problems raised by the fact that we have two cerebral hemispheres, each with an immense amount of localized performance, with inputs of general body sensation and vision channelled into the one or other side, and with movement likewise dependent on the one or the other side of the motor cortex; yet we experience what BREMER (1966) aptly calls a 'mental singleness.' We can ask: how can the diversity and the tremendous dispersion of activity in the spatio-temporal patterns of the brain give rise to this unity and, from moment to moment, the relative simplicity of our conscious experience so that the play of experience appears to be, as it were, all on the stage before one single conscious self?

Undoubtedly an essential neurological correlate of this unification of experience arising from neural events in the two cerebral hemispheres is the enormous commissural tract, the corpus callosum, that links the mirror-image areas of the two hemispheres; and to a lesser extent there is also the commissural linkage by the anterior commissure and the massa intermedia. As is well known, this mental unity in man remains intact after large lesions or surgical destructions of the cerebral hemispheres, even the interpretive areas (cf. Figs. 25, 30) for symbolic expression in language being destroyed; and we can all experience in dreams the fragmentary and chaotic imagery which is part of our experiencing unitary self. Perhaps even more remarkable are the observations of PENFIELD (1966, 1969), who was able to evoke remote audiovisual memory of illusory character by electrical stimulation of the temporal lobe; yet these strange experiences were assimilated by the subject and recognized as remembrances of long-forgotten incidents in the lifetime of the same self.

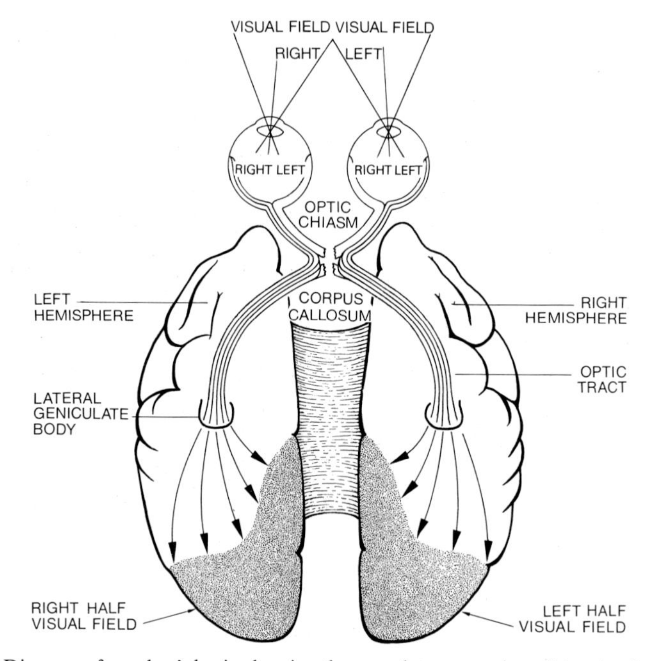

Fig. 23. Diagram of monkey's brain showing the complete separation of the visual pathways from the left and right eyes after sectioning the optic chiasma (modified from SPERRY, 1964)

The same extraordinary ability of the experiencing self to build a unity out of diversity is illustrated by the phenomenon of drug-induced hallucination. No matter how bizarre the experiences, they are recognized as belonging to the self and not due to some privileged view of mental happenings in some other self—or some amputated component of the original self, i.e. there is no mental diplopia.

This postulate of the key role of the brain commissures in linking the cerebral hemispheres has been tested in the last decade by MYERS and SPERRY and their associates (MYERS, 1961; SPERRY, 1964, 1966) in experiments of remarkable ingenuity. In the cat and monkey they have split the optic chiasma, so that each eye feeds into a cerebral hemisphere on its own side (Fig. 23). In the split-chiasma animals what was learned by visual inputs into one hemisphere was transferred to the other hemisphere and laid down as a memory trace, where it could be detected after a subsequent splitting of the brain; that is, the information that goes in from one eye to one hemisphere is communicated to the other hemisphere at the time of the laying down of memory traces. By contrast, after the brain

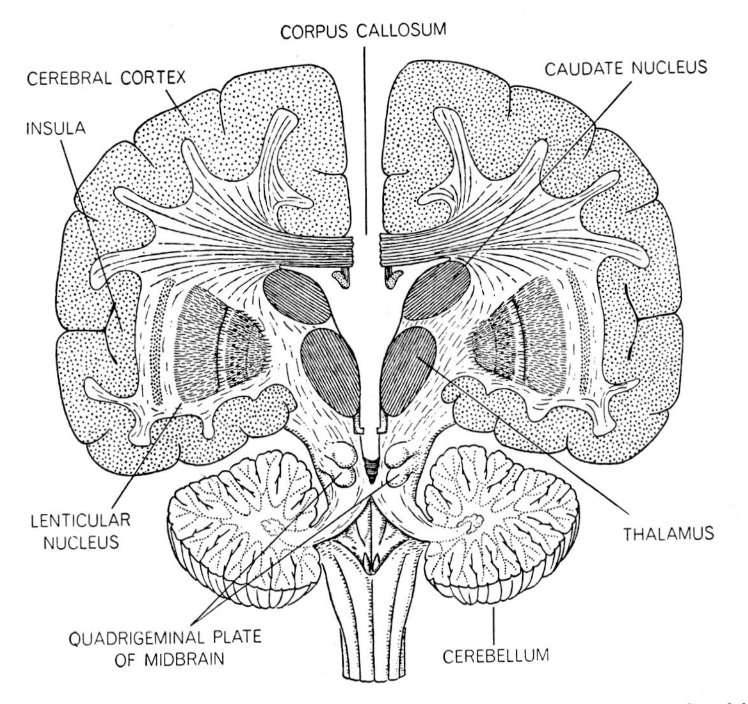

Fig. 24. Diagram showing the separation of the two cerebral hemispheres produced by section of the corpus callosum, which is shown as the great tract of interhemispheric communication. The diagram further shows a midline section of the cerebellum, but that is immaterial to the present discussion (SPERRY, 1964)

was split, there was no transfer of learning from one to the other side (Fig. 24). The two sides of the brain could be trained to give diametrically opposite responses to stimuli. At this level of testing, the animal is divided into two independently learning and behavioural organisms. By contrast, if this attempt to build up opposed responses to information fed into the two eyes of the split-chiasma animal was made before the splitting of the brain, a severe behavioural conflict was aroused in the animal. This behavioural conflict was also observed when attempts were made to train animals in an opposed manner with signals involving touch and kinesthesis. It can be concluded that the brain commissures are essentially concerned in the transfer of information between the two hemispheres, so that they can share in learning and memory. The neurological substrate of learning that is called an engram (LASHLEY, 1950), is normally laid down in both hemispheres of the cat. Experiments on monkeys and anthropoid apes reveal that this duality of memory trace is with them a less prominent feature. And with man there are remarkable examples of complex memories restricted to one hemisphere, as, for example, occurs

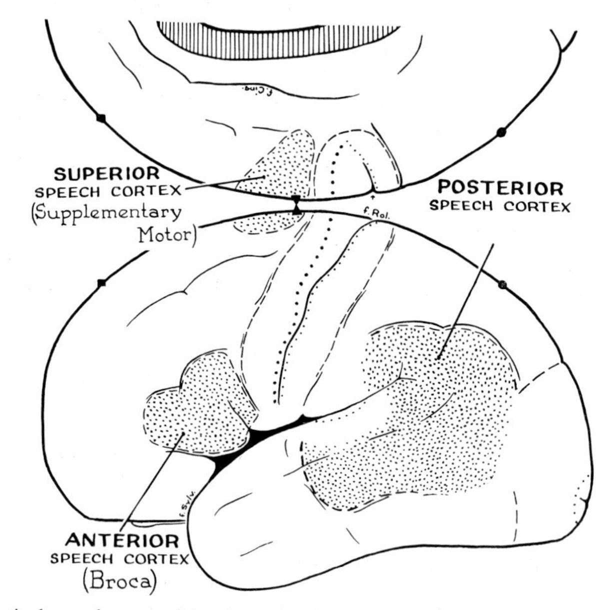

Fig. 25. Cortical speech areas of dominant hemisphere as determined by aphasic arrest by electrical interference (PENFIELD, 1966)

with language in the dominant hemisphere, which is the left in a right-handed subject (cf. Fig. 25).

The most remarkable examples of experiment on interhemispheric communication have been made on nine human subjects in which a surgical separation of the two cerebral hemispheres was made in order to control intractable epilepsy, and mercifully this operation of severing the corpus callosum, the anterior commisure and the massa intermedia was successful therapeutically (SPERRY, 1964, 1966). Just as with the split-brain animals, these subjects display no gross signs of incoordination of response, nor do they experience any splitting of their mental unity, such as might be called a "mental diplopia." However, they reveal gross disorders both of reaction and of experience when tested appropriately. The most remarkable findings stem from the almost invariable unilateral representation of language in the dominant cortical hemisphere (Figs. 25, 30), which is the left in all these cases. For example, they are unable to read with the left half of the visual field, which feeds exclusively into the right hemisphere (the minor hemisphere), and commands conveyed verbally are carried out with the right side only. They react to stimuli applied to the left visual field, sometimes appropriately, but without being able to give an account of what they are doing. Similarly, they have no detailed knowledge of touch or movements on the left side, and if blindfolded they do not know what the left side is doing. Evidently, the dominant hemisphere of the brain neither "knows" nor "remembers" the experiences and activities of the other hemisphere. The difficulties of these subjects derive from the uncontrollable behaviour of the left side, particularly the left hand. Often they try to control the left hand by their right hand.

All the evidence produced by these nine cases is explicable by the postulate that, when bereft of commissural linkages with the dominant hemisphere, the minor hemisphere behaves as a computer with inbuilt skills of movement, with recognition of the form and function of objects, and with the ability to learn; nevertheless, the dominant hemisphere with its ability of linguistic expression remains oblivious of all this performance. For example, as stated by SPERRY (1966),

"... the subject may be blindfolded and some familiar object like a pencil, a cigarette, a comb or a coin placed in the left hand. Under these conditions the mute hemisphere connected to the left hand feeling the object perceives and appears to know quite well what the object is. Though it cannot express this knowledge in speech or in writing, it can manipulate the object correctly, it can demonstrate how the object is supposed to be used, and it can remember the object and go out and retrieve it with the same hand from among an array of other objects either by touch or by sight. While all this is going on, the other hemisphere meanwhile has no conception of what the object is, and, if asked, says so. If pressed for an answer, the speech hemisphere can only resort to pure guess-work. This remains the case just so long as the blindfold is kept in place and all other avenues of sensory input from the object to the talking hemisphere are blocked. But let the right hand cross over and touch the test object in the left hand; or let the object itself touch the face or head as in demonstrating the use of a comb, a cigarette or glasses; or let the object make some give-away sound, like the jingle of a key case, then immediately the speech hemisphere also comes across with the correct answer."

Likewise, happenings in the left visual field remain unknown to the dominant hemisphere. For example, SPERRY (1966) reports:

"The subjects fail on such simple tasks, for example, as that involved in discriminating whether red and green half fields presented together are the same or different in colour where the response involves only a simple nodding or shaking of the head 'yes' or 'no'—in other words, with everything favouring any cross integration that might be present. The same task caused no difficulty to either hemisphere when the two colours or other stimuli were presented within the same half retinal field and hence projected to the same hemisphere. Comparison of the directional tilt of broad straight lines running across the visual field and interrupted in the centre of the screen went easily, again, when both parts of the bar fell in one field; but when the two parts fell in right and left fields separately, the subjects were unable to indicate whether the two bars were lined up straight across the midline or at an angle. When the response in this test involved manual copying of the perceived lines, the initial result was for each hand to record and draw only the part of the line within its own half of the visual field, the other half being omitted. When both hands had a pencil and worked simultaneously, both parts of the line were drawn correctly, indicating double simultaneous perception and response."

We can summarize this by stating that the goings-on in the minor hemisphere, which we may refer to as the computer, never come into the conscious experience of the subject.

It is particularly interesting to consider the problem of free-will in relation to these two separated cerebral hemispheres. So far as the dominant hemisphere and the right side of the body are concerned the situation is exactly as for a normal subject. However, the conscious subject has no direct control over what the left hand is doing, nor is there any cognizance of these movements except in so far as they produce information that is channelled by neural pathways into the dominant hemisphere, for example from the right visual field or with palpation by the right hand. We have already seen that the left hand can perform appropriate actions of recognition and that these are quite unknown to the conscious patient; in fact the dominant hemisphere may try, by using the right hand, to interfere with the correct responses being made by means of the minor hemisphere in response to information fed into it by the left visual field.

With progressive recovery from the operation, one of the subjects was able to develop some voluntary control of the left hand, but this was shown to be due entirely to the uncrossed pyramidal pathway from the left or dominant hemisphere directly to the nerve cells controlling the muscles of the left arm, and not at all to a conscious voluntary action exerted by the minor hemisphere, or computer. Yet there is plenty of evidence that this hemisphere does have a complex reacting life all of its own, though it has no power of overt expression because linguistic expression lies entirely with the dominant hemisphere. For example, by means of the minor hemisphere the left hand can point out a matching picture from many others that have been flashed into the left visual field. It can pick up the correct written name of an object flashed on the screen, or, *vice versa*, it can read a name and retrieve the designated object. For example, the words cup, fork or apple flashed on the screen cause the left hand to pick up the appropriate object that is also in the left visual field. All this occurs with the subject having no conscious experience whatever of what is happening with all these performances of the minor hemisphere, which in this way is displaying perception, comprehension, reading, retrieving and learning.

SPERRY (1966) argues that the presence of consciousness in a hemisphere would not be demonstrable in the absence of some appropriate linguistic mode of expression and therefore that the minor hemisphere may in fact be responsible for conscious states which cannot be indicated to the observer because of the failure of symbolic communication in language. In fact we can agree that the problem of trying to discover if the activities of the minor hemisphere are actually resulting in conscious experiences is equivalent to our problem of trying to discover if an animal's cerebral activity gives rise to conscious experiences. SPERRY even suggests that at some later date by more devious pathways, as for example the long-loop reticular pathways suggested by BREMER, it may be possible for the minor hemisphere to communicate with the dominant hemisphere and so be able to achieve linguistic expression. In that case it might even be able to report remembered experiences from the present experiments, though in the absence of present report it might be concluded that there were no such experiences!

An analogous situation can occur in the so-called automatic states of a subject with normal interhemispheric communication. Last century the great English neurologist, HUGLINGS JACKSON, made a special report of mental disorders which are now recognized as arising from localized epileptic seizures in the amygdaloid area. In these automatic states persons can, for quite long periods, react in a complex and sophisticated manner and yet, on recovery, have no trace of any recollection. For example, a doctor in an automatic state examined a patient and made reasonably accurate notes of the examination, yet remembered nothing afterwards. The question may be asked: was he conscious during this automatic state? My answer is that we cannot assume this to be so if the subject has no trace of any remembered experiences. In the absence of any evidence we must be agnostic, just as with the question of consciousness in animals.

Thus it is clear that the loss of commissural communication between the two halves of the brain has caused a split both in the perceptual and in the operational functioning of the person. The really remarkable finding is that the conscious self, with all its linguistic and sophisticated behavioural performance, seems to be represented solely in the dominant hemisphere in these split-brain patients. Unity of conscious experience is retained at the expense of a loss of all the experience that would be expected to be associated with the activities of the minor hemisphere. One may well wonder if this is the case in normal subjects, so that information fed into this hemisphere reaches consciousness only after interhemispheric transfer via the corpus callosum. This suggestion gives rise to the further question of what is the functional importance of the minor hemisphere other than to receive from the sense organs, to do complex computations thereon, then to transmit to the dominant hemisphere, and finally to receive from this hemisphere and transmit to the muscles of the opposite side. It should be mentioned, however, that motor skills and construction of spatial relations and perspective in drawing were much better performed by the left hand and the minor hemisphere using information from the left visual field.

In conclusion I feel a profound dissatisfaction when I contemplate the present attempts to account for the undoubted unity of my conscious experience. One is confronted by the extraordinary problem of trying to reconcile the unitary nature of my conscious self with the neurological events of the utmost diversity and complexity that are assumed to underlie it, and that involve the "weaving" by impulses of spatio-temporal patterns in the "enchanted loom" of SHERRINGTON with its thousands of millions of units or nerve cells. PENFIELD and JASPER (1954) attempt to soften this antithesis by postulating that the unification of experience occurs in the centrencephalic system which includes the great subcortical nuclei that give to and receive from all parts of the widespreading cerebral cortex. To them there is a relative simplicity in conscious experience in contrast to the enormous complexity of the sensory input. Yet I would argue that this simplification occurs only in respect of the moment-to-moment perceptions that we experience. By attention and concentration we can greatly sharpen the focus of experience and perceive a fineness of grain matching the information that the sense organs feed into the brain. In any case the problem of the unity of experience still remains an enigma whether the neural substratum is spatio-temporal patterns of neuronal activity in these large subcortical nuclei of the centrencephalic system, or in the cortex itself. As SHERRINGTON (1940) points out, there is no "centralization upon one pontifical nerve cell." The antithesis must remain that our brain is a democracy of ten thousand million nerve cells, yet it provides us with a unified experience.

Does the Uniqueness of the Experiencing Self Derive from Genetic Uniqueness?

We may take it as certain that my conscious self depends uniquely on my brain and not on other brains. I think telepathy is still a tenable belief; but, if it exists at all, it provides an extremely imperfect and inefficient way of transferring information from the neural activity of one brain and its associated conscious experiences to the conscious experiences arising from my brain. This unique interdependence between a brain and a conscious self raises a problem that has always been of great interest to me. It has been expressed by the great American biologist, H.S. JENNINGS (1930), in a speculative chapter entitled "Biology and Selves" in his book *The Biological Basis of Human Nature*. However, the climate of opinion has been so unfavourable to such speculations that Jennings's ideas have been almost universally neglected. Yet they are very relevant to the problems raised in this lecture: namely, the uniqueness of the conscious experiences that each of us enjoys, and their relationship to the neuronal activities of our brains.

Two questions may be asked: what is the nature of this consciously experiencing self? and how does it come to be related in this unique manner with a particular brain? I am aware that to many these are not valid questions. My only rejoinder can be that to me they are the most fundamental and important questions that can be asked; and let me state, as a brief autobiographical aside, that I have held this belief since I was 18 years old, when I had a kind of sudden illumination of these problems, and I have been driven on by their interest and urgency to spend my life studying the nervous system.

JENNINGS formulated with a masterly and lucid style two problems that to him were quite unanswerable. Both were related to the superficially attractive hypothesis that the uniqueness of the self derives from the uniqueness of the particular gene combination belonging to that self, or, as JENNINGS expresses it, "the assumption that it is diversity of gene combination that gives origin to distinctiveness of selves."

In the first place, of course, that assumption is refuted by the distinctiveness that is experienced by identical twins with their identical gene combinations. Alike as these twins are to external observers, each in its own conscious experiences and self-hood is as distinct from its fellow twin as it is from any other self. Evidently, identity of gene combinations must be compatible with distinctiveness of experiencing selves.

The second problem has a universal reference to all conscious selves, to each one of us. It was formulated by JENNINGS in relation to the genetic theory that any individual (except identical twins) genetically is a unique and never-to-be-repeated knot of strands of genes (DOBZHANSKY, 1962, 1967) that has come by inheritance through countless individuals from the remote past. JENNINGS asks:

"What is the relation of my self, identified as it is with one particular knot in the great network that constitutes humanity, to the other knots now existing? Why should I be identified with one only? To an observer standing apart from the net, it will not appear surprising that the different knots, since they are formed of diverse combinations of strands, should have different pecularities, different characteristics. But that the observer himself—his total possibility of experience, that without which the universe for him would be non-existent—that he himself should be tied in relations of identity to a single one of the millions of knots in the net of strands that have come down from the unbeginning past—this to the observer appears astonishing, perplexing. Through the operation of what determining causes is my self, my entire possibility of experiencing the universe, bound to this particular one of the combination of strands, to the exclusion of some millions of others? Would *I* never have been, would *I* have lost my chance to participate in experience, would the universe never have existed for me, if this particular combination had not been made?

If the existence of *me* is thus tied to the formation of a particular combination of genes, one may enter upon calculations as to the chances that I should ever have existed. What are the chances that the combination which produced me should ever have been made? If we depend on the occurrence of the exact combination of genes that as a matter of fact produced us, the odds are practically infinite against your existence or my existence.

And what about the selves that would have come into existence if other combinations of genes had been made? If each diverse combination yields a different *self*, then there existed in the two parents the potentialities, the actual beginnings, of thousands of billions of selves, of personalities, as distinct as you and I. Each of these existed in a form as real as your existence and my existence before our component germ cells have united. Of these thousands of billions, but four or five come to fruition. What has become of the others?"

And of course to go further backwards in our genetical tree makes the problem even more preposterously fantastic. Hence on both these grounds, I must reject this materialistic doctrine that the uniqueness of my conscious experiencing self is derived from the uniqueness of my genetic make-up. What then determines the uniqueness of my self?

I have found that a frequent and superficially plausible answer to this question is the statement that the determining factor is the uniqueness of the accumulated experiences of a self throughout its lifetime. And this factor is also invoked to account for the distinctiveness of uniovular twins despite their genetic identity. It is readily agreed that my behaviour, my character, my memories, and in fact the whole content of my inner conscious life are dependent on the accumulated experiences of my life; but no matter how extreme the change that can be produced by the exigencies of circumstance, I would still be the same self able to trace back my continuity in memory to my earliest remembrances at the age of one year or so, the same self in a quite other guise. Thus the accumulated experiences of a lifetime cannot be invoked as the determining or generating factor of the unique self, though naturally they will enormously modify all the qualities and features of that self. The situation is somewhat analogous to the Aristotelian classification into substance and accidents.

JENNINGS must have appreciated the fallacy of attempting to derive the uniqueness of self from the experiential history of an individual, for in searching for an explanation he develops the following remarkable speculations:

"To work this out in detail, one would apparently have to hold that the human self is an entity existing independently of genes and gene combinations; and that it merely enters at times into relations with one of the knots formed by the living web. If one particular combination or knot should not occur, it would enter into another. Then each of us might have existed with quite different characteristics from those we have—as our characteristics would indeed be different if we had lived under different environments .... It could be held that there is a limited store of selves ready to play their part, that the mere occurrence of two particular cells which may or may not unite has no determining value for the existence of these selves, but merely furnishes a substratum to which for reasons unknown they may become temporarily attached .... And what interesting corollaries might be drawn from such a doctrine, as to the further independent existence of the selves after the dispersal of the gene combinations to which they had been attached! Certainly no one can claim that biological science establishes or indeed favours that doctrine. But since biology itself furnishes no positive doctrine of the relation of selves to gene combinations, the question is a fair one: Does biological science make the holding of that doctrine impossible?"

General Conclusions

It is important to recognize that in the first instance this question of the relation of a self to gene combinations can be asked only by an experiencing self of its own existence. For example, I can ask it in relation to my own self, and I reply that I must face up to the problems of my own personal existence as an experiencing self that is dependent on the functioning of a brain, which I try to understand as a biological mechanism; and that my brain has had a biological origin as a consequence of a gene combination and the ensuing embryological development. My experiencing self is the only reality I know by direct apprehension—all else is a second-order or derivative reality. The arguments presented by JENNINGS preclude me from believing that my experiencing self has an existence that merely is derivative from my brain with its biological origin, and with its development under instructions derived from my genetic inheritance. Nor do I believe with the physicalists that my conscious experiences are *nothing but* the operation of the physiological mechanisms of my brain. It may be noted in passing that this extraordinary belief cannot be accommodated to the fact that only a minute amount of cortical activity finds expression in conscious experience. Contrary to this physicalist creed, I believe that the prime reality of my experiencing self cannot with propriety be *identified* with some aspects of its experiences and its imaginings—such as brains and neurones and nerve impulses and even complex spatio-temporal patterns of impulses. The evidence presented in this lecture shows that these events in the material world are necessary but not sufficient causes for conscious experiences and for my consciously experiencing self.

If we follow JENNINGS, as I do, in his arguments and inferences, we come to the religious concept of the soul and its special creation by God. I believe that there is a fundamental mystery in my existence, transcending any biological account of the development of my body (including my brain) with its genetic inheritance and its evolutionary origin; and, that being so, I must believe similarly for each one of you and for every human being. And just as I cannot give a scientific account for my origin—I woke up in life, as it were, to find myself existing as an embodied self with this body and brain—so I cannot believe that this wonderful divine gift of a conscious existence has no further future, no possibility of another existence under some other unimaginable conditions. At least I would maintain that this possibility of a future existence cannot be denied on scientific grounds. This theme is further developed in Chapter X.

For a final statement of my belief, I would like to quote from an earlier EDDINGTON lecture by THORPE (1961):

"I see science as a supremely religious activity but clearly incomplete in itself. I see also the absolute necessity for belief in a spiritual world which is interpenetrating with and yet transcending what we see as the material world .... Similarly I believe that anyone who denies the validity of the scientific approach within its sphere is denying the great revelation of God to this day and age. To my mind, then, any rational system of belief involves the conviction that the creative and sustaining spirit of God may be everywhere present and active; indeed I believe that all aspects of the universe, all kinds of experience, may be sacramental in the true meaning of the term."

Chapter VI

Evolution and the Conscious Self[1]

Introduction

We are participants in the evolutionary process of life, yet only in the last 100 years has man realized his evolutionary origin. The full implications of this re-orientation of man to nature have not yet been lived with long enough for them to be assimilated to man's conceptual thinking about himself. The emotional controversies of last century continued into this century and have delayed a wise evaluation of the evolutionary story in relation to man. However, in the last few years there have been publications by leading biologists (DOBZHANSKY, 1962, 1967; SIMPSON, 1964; LACK, 1961; THORPE, 1962) that reveal the beginning of an evolutionary philosophy based upon a mature understanding of the evolutionary process as it is at present formulated.

When we consider the story of the evolutionary development of living forms, we tend to regard ourselves as being onlookers of this evolutionary procession, as we try to hold in thought the immensity and wonderful productivity of this biological process. But we are *in* the procession. It is not enough for us to think of man in general as being so engaged. It is the sense of personal involvement that we must realize emotionally. Each of us, I and you, is at the end of a line of genetic descent that stems from the earliest living organisms. The vision I have is that of a great procession moving through time. We can sense directly only the contemporaneous living forms that, as it were, are stretched out across the line of the procession. We can guess and imagine and reconstruct what went on before by what has been dropped by our predecessors—the writings, the art, the artefacts, and then just the skeletal remains—as in

1 This is the text of a lecture delivered to a large student audience on January 11th, 1967 at Gustavus Adolphus College, St. Peter, Minnesota. This Third Annual Nobel Conference was on the theme "The Human Mind." The original lecture form of the text has been retained and the text has only been slightly modified from the Conference publication (ECCLES, 1967).

imagination we reconstruct from such traces earlier and earlier pictures of what has gone on before us. And what will be following us lies in impenetrable darkness. It is as if we are moving in the opposite direction from the way that we are facing, so that we are looking backwards and can see only the way we have come.

We are in this evolutionary procession several thousand million years after its beginning. Of course, the temptation is great to assume a God-like attitude and tacitly to view evolution from the outside—or to be impersonal and talk of mankind in general as being a product of the evolutionary process. But for the realization of man's true position we must recognize our personal position in this procession of biological creation—our very personal existence was dependent on all the exigencies throughout this immense time of the evolutionary process. There are various reasons why we are embarrassed or fearful of such engagement. It is so much more reassuring to have the detachment of a God-like onlooker at life and at all the rest of mankind.

The most marvellous theme of the evolutionary story is the development of the nervous system. Amidst all the perfecting of this or that biological form to make it more fitted to survive, nothing can compare with the pre-eminent role of the nervous system, that organizes the whole animal and gives it a unity in its reaction to the environment. Informed about the external environment by sense organs and with muscles at its command, the nervous system controls the life of all but the most primitive animals.

Initially, the functioning of the nervous system would be dependent solely on the structure deriving from the inborn genetic inheritance; but even the higher species of invertebrates, such as the octopus, can be readily trained or conditioned, so that the reaction of an individual to its environment comes to be a product not only of inheritance, but also of the encounters during its life-time (YOUNG, 1964); and there is evidence that even much simpler organisms, such as worms, have this faculty.

The story of vertebrate development to the eventual mastery of this planet is essentially the story of neural development. Let us appreciate that all skills in reaction—the effortless control of flight by birds, the gymnastics of monkeys, the grace of felines, the virtuosity of a ballerina, or of a pianist or of a violinist, the skill in games or in craftsmanship—all are splendid manifestations of the nervous system in its control of behaviour. All sense organs—eye, ear and the rest—are but signal systems giving significant information to the brain; and all muscular movement is but the carrying out of orders transmitted from the central nervous system.

So the story of vertebrate success in dominating not only the land, but also the air and the sea, was witness to the neural mastery even before man's origin. And this story of neural mastery culminates in man, so much so that man by planned action now controls even the evolutionary process.

The Modern Theory of Evolution

Essentially the modern theory of evolution has two components. There are firstly the changes being wrought in the genetic code by random mutations and recombinations which change the DNA structure of the sex cells of an organism. The vast majority of these mutations are detrimental to the organism developing from such a sex cell, and such mutations are consequently rapidly eliminated. However, the rare cases of a mutation favourable for survival and multiplication will lead to a flourishing of that organism and its descendents bearing the same favourable genetic code. This illustrates the other operative component in evolutionary theory. It is called "natural selection," and it is believed to be responsible for the marvellous adaptations developed in the evolutionary process. TEILHARD DE CHARDIN (1959) has likened the evolutionary process to a "blind groping." The randomness of the mutational changes gives essentially a blind trial of an enormous number of possible changes, while natural selection is responsible for the survival of all "favourable" mutations. Hence from an initial process of pure chance there can be wrought by natural selection all the marvellous structural and functional features of living organisms with their amazing adaptiveness and inventiveness. As so formulated, the evolutionary theory is purely a biological process involving mechanisms of operation that are now well understood in principle, and it has deservedly won acceptance as providing a satisfactory explanation of the development of all living forms from some single extremely simple primordial form of life. This theory stemming from DARWIN must rank as one of the grandest conceptual achievements of man.

Transcendences in the Evolutionary Story

There would thus seem to be a continuum in the evolutionary process, since it provides an explanation for the whole world of biology. However, as has been well stated by DOBZHANSKY (1967), there were two exceptions: the origin of life and the origin of man, the latter being the special theme of this discourse.

"The origin of life and the origin of man were evolutionary crises, turning points, actualizations of novel forms of being. These radical innovations can be described as transcendences in the evolutionary process. Human mind did not arise from some kind of rudimentary 'minds' of molecules and atoms. Evolution is not simply unpacking of what was there in a hidden state from the beginning. It is a source of novelty, of forms of being which did not occur at all in the ancestral states. Here inevitably arises the thorny question: were these new forms of being determined to appear? And if so, were they real novelties?

This should not be taken to mean that the origin of life or of man were instantaneous or even very swift. A process which is very rapid in a geological (more precisely, paleontological) sense may appear to be lengthy and slow in terms of a human lifetime or a generation. The appearance of life and of man were the two fateful transcendences which marked the beginnings of new evolutionary eras."

There is at present much discussion and even scientific experimentation related to the first transcendence, the problem of the origin of life. It is important, at the outset, to discuss what we mean by life, and I would suggest that "life as we know it" is the simplest definition. This life would be carbon-based and exist in watery media, with such fundamental organic compounds as amino acids, proteins, nucleic acids and lipids. "Life as we do not know it" could be based upon some other elements, such as silicon instead of carbon, ammonia instead of water and so on; but it is important to recognize that there is no evidence whatsoever for an organic existence on such a basis, and it has not even been shown that there is any plausibility about such a suggestion. In any case, if we were to discover the existence of "life as we do not know it," it would probably be classifiable as a third kind of existence of a different order from either our living or non-living categories.

In some remarkable experiments in recent years MILLER (1955, 1957) and CALVIN (1961, 1969) and their associates have shown that, if the earth contained a primitive atmosphere of gases rich in hydrogen, such as methane, ammonia and water, there is a synthesis of many types of organic compounds when such an atmosphere is subjected to spark discharge, ultra-violet radiation or cosmic radiation. For example, amino acids, adenine, acetic acid and succinic acid have been synthesized in this way, and there are suggestions about how these simple organic substances could be stabilised for further development, such as the production of macromolecules by further synthesis and by polymerization (FOX, 1964).

This work certainly is of great interest in connection with the origin of life, but unfortunately the paleogeologist PHILIP ABELSON (1956) has questioned on geological grounds the composition that is assumed for the primitive atmosphere. Thus the whole question of the origin of the simplest organic compounds is still uncertain.

But even if these elementary organic compounds were formed, as postulated by MILLER (1955, 1957) and CALVIN (1961, 1969), it is hard to see how they could accumulate to any appreciable degree [2]. Before anything

2 Recently KATCHALSKY (personal communication) has shown that certain clays are very effective in concentrating and stabilizing such organic compounds. FOX (1964) has suggested that concentration may occur by evaporation of isolated pools, and in model experiments has shown that polymeric material may become organized into "microspheres."

approaching the most primitive living organism can arise, there must be at least four major developments after the stage of production of the essential organic compounds. Firstly, there must somehow be an organization of a cellular type, encapsulated by a surrounding membrane. Secondly, there must be some chemical process within this cell providing the mechanism for building up an organization of energy-rich substances. Thirdly, the surrounding membrane must be permeable to substances so as to allow exchange with the environment and the building up of its contents. And finally, there must be some mechanism within this cell for storing and replicating information that it receives from its environment.

This last property is essential in order that evolution can occur, and so give rise to all the diversity of living forms. There are two essential requirements for evolution. One is that the primitive organism can acquire new information and store it and transmit it to its progeny. For example, this is the genetical, mutational aspect of evolution. Secondly, there must be some kind of feedback and encoding of information that leads to more diversification of the organism and a better adaptation to its environment.

As G. G. SIMPSON (1964) states in his book, "This View of Life: The World of an Evolutionist":

"These processes of adaptation in populations are decidedly different in degree from any involved in the prior inorganic synthesis of macromolecules. They also seem to be quite different in kind. It can be realized that much more is required in the development of even the simplest living organism than simply a creation of organic complex molecules. One must appreciate that the living organism transcends the simple physical and chemical properties of its elementary constituents, requiring some holistic development that characterises even the simplest living forms."

We can conclude, therefore, that, though there is likely to be a wide variety in production of organic compounds, there is a high degree of improbability that from these organic macromolecules a true cellular life could develop. Nevertheless, if we have, for example, one hundred million planets, as estimated by HARLOW SHAPLEY, that provide physical conditions suitable for life, it is conceivable that life could have arisen many times in the galaxy.

The most remarkable feature of "life as we know it" is that, in spite of its extreme diversification over the whole range of animals and plants, the same DNA code with the four so-called letters of the genetic alphabet, ATGC, is shared by all.

"All biological evolution, extending over a period of some two billion years, has occurred on the level of genetic 'words' and 'sentences,' no new 'letters' having been added or, as far as known, lost. The simplest interpretation of this is either that life arose only once and all living things stem from this one event, or that the existing 'genetic alphabet' proved to be more efficient than the others and is the only one which endured' (DOBZHANSKY, 1962)."

It would appear that the "origin of life" is an event of fantastic improbability, which of course is in conformity with its classification as a "transcendence," and which is in contrast to the extremely fertile creativity of the ordinary operations of the evolutionary process with the millions of species at present existing, and the immense number of extinct species.

As DOBZHANSKY clearly states in the earlier quotation, it is the origin of the human mind that represents a second transcendence in the evolutionary process. There is no special problem in regard to man's body as such. There is now no problem of the "missing link" in the evolutionary chain. The general sequences of man's origin are now well documented and the time scale is fairly well known. Hence it is often asserted that there are no fundamental problems outstanding with respect to the evolutionary origin of man.

Of course, I accept this explanation of my own origin. I cannot doubt my animal ancestry, and I regard as well established the biological mechanisms of evolution—mutations and natural selection. Yet I do not believe that this theory provides a complete explanation of my origin. I can believe that, so far as the human body (my body) is concerned, the evolutionary theory gives a fairly adequate account. However, this theory fails completely to provide me with an explanation of my origin as the person I experience myself to be with my self-awareness and unique individuality, to which Chapters IV and V have been devoted.

Human Self-Awareness or Conscious Experience

The unique human character of self-awareness has been well expressed by FROMM (1964):

> "Man has intelligence, like other animals, which permits him to use thought processes for the attainment of immediate, practical aims; but man has another mental quality which the animal lacks. He is aware of himself, of his past and of his future, which is death; of his smallness and powerlessness; he is aware of others as others—as friends, enemies, or as strangers. Man transcends all other life because he is, for the first time, life aware of itself. Man is in nature, subject to its dictates and accidents, yet he transcends nature because he lacks the unawareness which makes the animal a part of nature—as one with it."

It has been well-said that an animal knows, but only a man knows that he knows. As DOBZHANSKY (1967) comments:

> "Self-awareness is, then, one of the fundamental, possibly the most fundamental, characteristic of the human species. This characteristic is an evolutionary novelty; the biological species from which mankind has descended had only rudiments of self-awareness, or perhaps lacked it altogether. The self-awareness has, however, brought in its train somber companions—fear, anxiety, and death-awareness."

In order to preserve a continuity in the evolutionary process and to avoid a special and unique emergence or discontinuity—the second transcendence postulated by DOBZHANSKY (1967)—many eminent thinkers (SHERRINGTON, 1940; TEILHARD DE CHARDIN, 1959; HUXLEY, 1962) have taken refuge in the vague generalization that there is a mental attribute in all matter. As the organization of matter gradually became perfected in the evolutionary process, there was a parallel development of the mental attribute from its extremely primordial state in inorganic matter, or in the simplest living forms, through successive stages until it reached full fruition in the human brain. This statement is often expressed as if it were scientifically established, which is certainly not true. It is a purely gratuitous assumption that inorganic matter or that the simplest organism has some mental attribute that is refined and developed in the evolutionary process. I do not understand in what sense the words "mind" and "mental" are used in such statements. Evidences of reactions to stimuli and of apparently purposive movements in animals are naively regarded as establishing that they have conscious experiences of the same nature as those that you or I for example directly experience, and which can only be directly known to each one of us. We must remain agnostic with regard to the consciousness or self-awareness of animals.

Such statements of a progressive emergence of conscious mind during evolution are not supported by any scientific evidence, but merely are statements made within the framework of a faith that evolutionary theory, as it now is, will at least in principle explain fully the origin and development of all living forms including ourselves. In fact, the only evidence that we have is against the belief that there is a mental attribute in all matter, even the most highly organized matter, which occurs in the brain. Our conscious experience arises in only one part of our body—the highest levels of the brain—and even then only when the brain is in the right state of dynamic activity (Chapters IV, V; ADRIAN, 1947). Sentience of any part of my body is dependent on its functional nervous connection with my brain. Interruption of nervous connection between any part of the body and the brain, makes that part anaesthetic. And more remarkable still are the recent observations of SPERRY (1966) on humans with brains split for therapeutic purposes. Under such conditions all the goings-on in the minor cerebral hemisphere are not experienced (cf. Chapter V).

It is relevant to quote the excellent summary of LACK (1961) on the evolutionary origin of man.

"It seems certain that man's physical evolution from apelike ancestors occurred gradually by natural means, and it is reasonable to postulate the same for his overt behavior, including his social behavior. But, though it would seem to follow that man's inner attributes, including his moral conduct, likewise arose by gradual and natural means, the attempts to bridge the apparent gap between animals and man in this respect have been highly unconvincing. On the one hand, arguing from man downward, the upholders of Creative Evolution have postulated a 'Life Force' directing the mutations, a 'psyche' in the molecules, or 'goodness' or 'mind' in the stuff from which the universe is made, which are concepts of teleology or metaphysics unobservable by science. On the other hand, arguing from animals upward, the upholders of Evolutionary Ethics have postulated that moral conduct is a product of natural selection, thus reducing it to social behavior and missing the essence of the human experience; and this involves bringing a scientific concept into a branch of philosophy where it is at best irrelevant. Both these types of approach should be rejected, not through inadequate evidence, but because each involves the extension of a branch of learning beyond its proper terms of reference."

"TEILHARD DE CHARDIN (1959) has a strong case for asserting that an intermediate stage is inconceivable between the absence and the possession of self-awareness, and the same might be argued of free-will and moral conduct (even though, once acquired, they might be developed further). Nevertheless, it offends one's preconceptions of economy to postulate a break in evolutionary continuity, even if solely with respect to man's inner world of values. Perhaps the difficulty arises because it is wrong to picture the problem as a 'gap' to be 'bridged.' We readily accept the idea of a bridge betwen ape-like forms and man with respect to anatomy and overt behavior, because other animals show at least rudiments of what is found in man, so the bridge is securely founded at both ends. But whether anything of man's moral conduct and other inner attributes is present in other animals, and if so in what form, is unobservable, so in respect to these attributes there is no foundation for a bridge at the animal end."

The Evolutionary Story of Man's Origin

During the last decades much of the story of man's origin has been illuminated and we now have the main outlines of the greatest of all the happenings on this earth of ours. One of the most amazing discoveries has been that the evolution of man occurred over such a large area of the earth—South Africa (Transvaal), Tanganyika, China (Pekin), Java, and Europe. Both the characteristic features of evolution are represented: cladogenesis, the multiplication of species and the consequent diversification; and anagenesis, the perfecting of a single species. However, as emphasized by DOBZHANSKY (1962, 1967), it appears that, at any one place and time, there was only one species of hominids or primitive men, i.e. a single breeding community developing by anagenesis.

The earliest ancestors of man *(Australopithecus)* that are clearly distinguishable from the apes lived about one million years ago in Africa (Tanganyika and the Transvaal). They walked erect and had primitive pebble tools, but their brains, though large (500 cc), were still comparable in size with those of apes, both then and now. There is doubt as to whether these progenitors of man should be classified as apes *(Pongidae)* (SIMPSON, 1945) or as *Homo* (HEBERER, 1959). However, the recent discoveries in Tanganyika have revealed that much later there was a progenitor of man *(Homo habilis)* that is clearly classified as belonging to the genus *Homo*, and that probably arose from one of the species of *Australopithecus*.

Beginning about 500,000 years ago there were several races of the species, *Homo erectus*, scattered over Java, China, Africa, and Europe. The domestication of fire was first accomplished in China where *Homo erectus* had achieved the amazing advance in brain capacity to 900 to 1,200 cc. There are various isolated examples of further developments of brain and performance, but the most remarkable came only about 100,000 years ago during the last glaciation, when the Neanderthal race of *Homo sapiens* was living in Europe and West Asia. They had fire and primitive stone tools and possibly used animal skins for clothing. Here we have at last a race of men whose brain capacity (1400 cc) was equivalent to that of modern man and who had already developed what we may call a rudimentary spirituality, because they buried their dead.

I would suggest that ceremonial burial customs provide us with the first evidence for the dawn of self-consciousness in the developing hominids during the evolutionary process, perhaps some hundred thousand years ago. We can imagine that these customs developed when in some primitive way there arose the idea that the living person was more than a body, but was indeed a conscious or spiritual being.

It has been suggested that the harsh conditions prevailing during the last glaciation were responsible for the great evolutionary advance, but a surprising feature is that, with the improving climatic conditions 35,000 to 40,000 years ago, a new race of men, *Homo sapiens sapiens*, replaced Neanderthal man and rapidly wandered over the whole earth, Europe, Asia, Africa, America, Australia. The whole family of man today belongs to this wandering colonizing race. Unfortunately, the site and mode of origin of this last race of man is still not sufficiently documented, but they seem to have arisen outside Europe and were quite different in their tool culture from Neanderthal man.

Criteria for Self-Consciousness or Self-Awareness

After this survey of the evolutionary origin of man, it is necessary to examine the criteria we would use in determining if possible whether a particular hominid had self-awareness or not. It is generally agreed now that burial customs provide us with far the best criterion of self-awareness. Let us first remember that no animal in the wild state displays any concern for its dead and as a consequence there is not even a primitive attempt at any effort to dispose of the dead body, which is just ignored. Admittedly more information would be desirable from investigators such as Jane Goodall about anthropoid apes living in the wild state and their reactions to death and the dead bodies of their companions. Necessarily, domesticated animals must be excluded because so much of their behaviour is imitative of humans.

When we come to the evidence of burial customs among primitive men, it is conclusive only at the level of Neanderthal man. The most ancient known burial is in Palestine where several bodies were laid in graves cut in the floor of the cave and accompanied by offerings of food and weapons. There are many other examples of Neanderthal graves with similar ceremonial burials. As man moved from Paleolithic and Mesolithic to Neolithic times, there was progressive development in burial customs. I will give two quotations from DOBZHANSKY (1967) on these most important findings.

> "The concern for the dead became widespread evidently at the dawn of humanity. It continues to be one of the cultural universals in mankind. The forms it takes in different cultures are, however, widely variable, ranging all the way from fear and dread of the spirits of the deceased, to invoking them as helpers and protectors. The bodies are buried under the floors of the dwellings in which their kin live, or in special cemeteries at a distance from the dwellings, or in natural or artificially made caves or tombs. Or else, the corpse is cremated, and the ashes preserved or buried in special containers or scattered or thrown into water. Care may be taken to preserve as much as possible the outer appearance of the deceased by embalming or mummification, or the body may be deliberately exposed to wild animals or to carrion vultures .... The cardinal fact is that all people everywhere take care of their dead in some fashion, while no animal does anything of the sort."
>
> "Only a being who knows that he himself will die is likely to be really concerned about the death of others. JOHN DONNE's genuinely great insight: 'Any man's death diminishes me, because I am involved in mankind' is not spoiled even by having become a cliché. DONNE has perceived that it is chiefly because of my death-awareness that I feel myself 'involved in Mankinde.' Anthropologists have recorded a great variety of customs and attitudes to death and its sequels in different peoples. All these customs are grounded on the fundamental fact of death-awareness, which is indeed one of the basic characteristics of mankind as a biological species."

It, therefore, can be safely inferred that Neanderthal man had certainly self-awareness of the kind experienced by us and had the feeling for other members of his community as beings like himself. Thus perhaps 100,000 years ago we could speak of man's ancestors as being at the dawn of humanity and to have become self-conscious beings. Already at that time they had brains not inferior to ours in size, though the Neanderthal man would hardly be a prepossessing person, according to our standards, with his short, stooped and stocky frame.

The next great advance that we observe in man is provided by the artefacts that have been wonderfully preserved in the cave paintings and carvings of Southern France and Northern Spain. Here perhaps 20,000 years ago were highly civilized people capable of symbolic thought and representation and with cultivated skills in craftmanship. Since their pictorial art is far superior to that of primitive people today, it can safely be assumed that they had a highly developed language, though of course only in the spoken form. The first ideographic expressions of language would be still far in the future.

It is language that gives man the immense superiority over animals because it opens up the limitless possibilities of using word symbols for things, and so holding objects in memory when no longer observable; and then of course being able to develop more and more sophisticated usages of language in higher levels of symbolic and abstract thought. These potentialities are shared by all members of the human family, and all have their own highly developed languages, though in other respects their culture may be exceedingly primitive. As SUZANNE LANGER (1951) states:

"If we find no prototype of speech in the highest animals, and man will not say even the first word by instinct, then how did all his tribes acquire their various languages? Who began the art which now we all have to learn? And why is it not restricted to the cultured races, but possessed by every primitive family, from darkest Africa to the loneliness of the polar ice? Even the simplest of practical arts, such as clothing, cooking, or pottery, is found wanting in one human group or another, or at least found to be very rudimentary. Language is neither absent nor archaic in any of them."

In supporting the belief that there is in general an equivalence of brain potentiality in all races, I can think of no more striking statement than that of SUZANNE LANGER (1951):

"Language is, without a doubt, the most momentous and at the same time the most mysterious product of the human mind. Between the clearest animal call of love or warning or anger, and a man's least, trivial word, there lies a whole day of Creation—or, in modern phrase, a whole chapter of evolution. In language we have the free, accomplished use of symbolism, the record of articulate conceptual thinking; without language there seems to be nothing like explicit thought whatever. All races of men—even the scattered, primitive denizens of the deep jungle, the brutish cannibals who have lived for centuries on world-removed islands—have their complete and articulate language. There seem to be no simple, amorphous, or imperfect languages, such as one would naturally expect to find in conjunction with the lowest cultures. People who have not invented textiles, who live under roofs of pleated branches need no privacy and mind no filth and roast their enemies for dinner ... will yet converse over their bestial feasts ... in a tongue as grammatical as Greek, and as fluent as French! Animals, on the other hand, are one and all without speech .... Careful studies have been made of the sounds they emit, but all systematic observes agree that none of these are denotative, i.e., none of them are rudimentary words."

Not only have the anthropoid apes no language of their own, but they cannot be taught to speak a human language. There is a moving example of this in the story of the KELLOGGS who reared a chimpanzee baby, Gua, with their own child (KELLOGG and KELLOGG, 1933). They found that, though living in an environment of human speech, Gua made no effort to learn new movements of the muscles used in vocalization, so that she continued only with a few sounds indicative of provocation; whereas the child was continuously engaged in vocal play and learned several words during the nine months of the experiment. Other attempts to teach apes language have similarly been unsuccessful, though after immense labour for six months W.H. FURNESS (1916) succeeded in teaching a young orang-utan to use two words, papa and cup. But despite five years of effort he failed to teach a bright chimpanzee to do more than to pronounce one word "mamma" very poorly! The inability of a chimpanzee to learn to speak has also been reported by HAYES and HAYES

pre-Copernican, because surely there is a much more important criterion than some trivial Newtonian mechanics, which has no significance in the mechanics of relativity theory. I suggest that this criterion is that the earth is man's home and that but for us, with our intelligent conscious and scientific life, the whole drama of the cosmos might never be perceived and understood as it is in modern astronomy. As SCHRÖDINGER so poignantly surmises, it would be no more than a drama played before empty stalls!

General Considerations

Because the uniqueness of the conscious self is not tied to some specific genetic inheritance, it cannot be assumed that it is a mere derivative of the creative process of evolution, though of course this origin pertains to the body *per se*. We need some radical rethinking in respect of our status and relationship to the evolutionary story. We are in it, but not exclusively of it. The perplexity of this problem arises because of our present inability to arrive at any satisfactory solution of the brain-mind problem (cf. Chapters IV, V and X).

It has been proposed that the problem of brain and mind can be resolved by assimilating it to Bohr's Principle of Complementarity. I personally am not satisfied by this ingenious manoeuvre, though it may point the way to a more radical solution. I seem dimly to apprehend the necessity for some revolutionary concept in psychophysiology that would result in a revolution at least the equivalent of the Theory of Relativity in Physics. That theory recognized that the observer is not merely passive in physical observations, but participates in respect both of the common frame of reference and of the interference that is necessarily produced in any system under observation. Similarly, as implied by SCHRÖDINGER (1958), it will be necessary in psycho-physiology to recognize that all observations initially are in a perceptual world of conscious experience, and that the external world of things and events is a derivative of the perceptual world. Further, it will be necessary to recognize that all our theorizing, as we construct scientific hypotheses or philosophical explanations, is also primarily an activity of the conscious mind, particularly in respect of its creative and critical abilities (cf. WIGNER, 1969). Inevitably there is some turning back on itself, as by conscious thought we try to gain insight into the working of the mind and its relation to the brain. The need for some psychophysiological theory of relativity so becomes imperative, and one can hope that the bright light of this anticipated theory will soon illuminate the present darkness.

We are in the evolutionary process of life, yet only in the last hundred years has man realized it. The full implications of this revolution in viewpoint have not yet been lived through and assimilated conceptually. The history so far has been that of a triumph of the evolutionists, against the anti-evolutionists—of action against reaction. Initially the implications and meaning of the evolutionary hypothesis were misunderstood on both sides. But now many biologists are realizing that a kind of pseudo-religion, Darwinism, is being foisted on us—to wit that we are in and of a cosmic process of evolution that in principle gives a complete explanation of our origin and our nature. You will realize that I am attacking Darwinism, not the scientific theory of evolution which I accept as a partial and limited explanation of my origin; but for me it fails as a complete and satisfactory explanation of my own personal existence. For me there is a profound mystery in existence. We cannot even anticipate any fundamental breakthrough in understanding; but at least we should have a far-ranging vision of the marvellous adventure we co-jointly find ourselves in—the adventure of life and in particular of the conscious life of the mind. This gives us all our civilization, our art as well as our science.

I would like to conclude by trying to re-assess the significance of the theory of evolution in relationship to the way in which man should think of himself. There is no doubt that in general mankind has felt a down-grading of man's status in this universe during the last three or four centuries. Firstly, our earth became a mere trivial component of the universe, and secondly man became merely a clever animal that arose by a biological process. I agree with DOBZHANSKY (1967) that this emotional devaluation of man is not justified by the scientific discoveries and arises from a misunderstanding of man's real position. As DOBZHANSKY said:

> "Evolution is a source of hope for man. To be sure, the modern evolutionism has not restored the earth to the position of the center of the universe. However, while the universe is surely not geocentric, it may conceivably be anthropocentric. Man, this mysterious product of the world's evolution, may be also its protagonist, and eventually its pilot. In any case, the world is not fixed, finished, and unchangeable. Everything in it is a product of evolutionary flux and development."

As a final illustration of my own beliefs I will quote from the Dead Sea Scrolls:

> "So I walk on uplands unbounded, and know that there is hope, for that which Thou didst mould out of dust to have consort with things eternal."

Chapter VII

The Understanding of Nature[1]

Science as Conjectures and Refutations

KARL POPPER has quoted from Xenophanes two passages which reveal how closely this pre-Socratic philosopher had anticipated POPPER'S views on the status of our attempts to understand nature—"a woven web of guesses."

"The Gods did not reveal, from the beginning,
All things to us; but in the course of time,
Through seeking, men find that which is better.

But as for certain truth, no man has known it,
Nor will he know it; neither of the gods,
Nor yet of all the things of which I speak.
And even if by chance he were to utter
The final truth, he would himself not know it;
For all is but a woven web of guesses."

(DIELS KRANZ: Fragmente der Vorsokratiker) (Trans. K. R. POPPER)

In the introduction to the Logic of Scientific Discovery (1959) POPPER makes a definitive statement of his philosophical position.

"I, however, believe that there is at least one scientific problem in which all thinking men are interested. It is the problem of cosmology: *the problem of understanding the world—including ourselves, and our knowledge, as part of the world.* All science is cosmology, I believe, and for me the interest of philosophy as well as of science lies solely in the contributions which they have made to it."

This is the most succinct statement that I know of my own beliefs and of the motivation of my own scientific efforts. We know of "the world—including ourselves" or "nature" solely by our conscious experiences—particularly those stemming from our sensory perceptions and their recall in memory. I assume that "understanding" means "comprehending these experiences by means of the explanatory power of scientific hypotheses of the utmost generality."

---

1 This chapter was originally written for a volume on the philosophy of KARL POPPER, and was later withdrawn, another contribution being written in its place.

POPPER made a fundamental contribution to the philosophy of the scientific method by his formulation of the problem of demarcation—or, otherwise expressed, of the criteria that establish whether or not a particular concept and its related observations are genuinely scientific. I cannot do better than to quote from POPPER (1963):

"Whenever a scientist claims that his theory is supported by experiment or observation, we should ask him the following question.

Can you describe any possible observations which, if they are actually made, would refute your theory? If you cannot, then your theory has clearly not the character of an empirical theory; for if all conceivable observations agree with your theory, then you are not entitled to claim of any particular observation that it gives empirical support to your theory.

Or to put it more briefly: only if you can tell me how your theory might be refuted, or falsified, can we accept your claim that your theory has the character of an empirical theory.

This criterion of demarcation between nonempirical theories and theories that have empirical character I have also called the criterion of falsifiability or the criterion of refutability. It does not imply that irrefutable theories are false. Nor does it imply that they are meaningless. But it does imply that, as long as we cannot describe what a possible refutation of a theory would be like, that theory is outside empirical science."

"Observations or experiments can be accepted as supporting a theory (or a hypothesis or a scientific assertion) only if these observations or experiments may be described as severe tests of the theory; or, in other words, only if they are the results of serious attempts to refute the theory, or of trying to fail it where failure should be expected in the light of all our knowledge, including our knowledge of competing theories."

A simple initial account of POPPER's philosophy of scientific method is given in the following quotation (1963),

"My thesis, as I have already indicated, is that we do not start from observations but always from problems—from practical problems or from a theory which has run into difficulties; that is to say, which has raised, and disappointed, some *expectations*.

Once we are faced with a problem, we proceed by two kinds of attempt: we attempt to guess, or to conjecture, a solution to our problem; and we attempt to criticize our usually somewhat feeble solutions. Sometimes a guess or a conjecture may withstand our criticism and our experimental tests for quite a time. But as a rule, we find that our conjectures can be refuted, or that they do not solve our problem, or that they solve it only in part; and we find that even the best solutions—those able to resist the most severe criticism of the most brilliant and ingenious minds—soon give rise to new difficulties, to new problems. Thus we may say that our knowledge grows as we proceed from old problems to new problems by means of *conjectures and refutations;* by the refutation of our theories or, more generally, of our *expectations*."

Science as a Personal Endeavour

It is generally believed that science has been created by the objective and impersonal activities of scientists. On the contrary it is one of the most personal of all human endeavours. I shall quote briefly again from MICHAEL POLANYI (1966), who has been the most eloquent and inspired writer on this theme (cf. POLANYI, 1958, 1967b, 1968a).

"To see a good problem is to see something hidden and yet accessible. This is done by integrating some raw experiences into clues pointing to a possible gap in our knowledge. To undertake a problem is to commit oneself to the belief that you can fill in this gap and make thereby a new contact with reality. Such a commitment must be passionate; a problem which does not worry us, that does not excite us, is not a problem; it does not exist. Evidence is cast up only by a surmise filled with its own particular hope and fervently bent on fulfilling this hope. Without such passionate commitment no supporting evidence will emerge, nor failure to find such evidence be felt; no conclusions will be drawn and tested—no quest will take place.

Thus the anticipatory powers that we have seen at work in historical perspective, arouse and guide individual creativity. These powers are ever at work in the scientist's mind, because he believes that science offers an aspect of reality and may therefore manifest its truth ever again by new surprises."

The degree of anthropomorphic involvement even in nuclear physics has been noted by MARTIN DEUTSCH (1959):

"In my own work I have been puzzled by the striking degree to which an experimenter's preconceived image of the process which he is investigating determines the outcome of his observations. The image to which I refer is the symbolic, anthropomorphic representation of the basically inconceivable atomic processes."

He further goes on to state that the creative scientific imagination functions by evoking potential or imagined sense impressions. As GERALD HOLTON (1967) comments:

"The more carefully we peer at the 'faces' of our meters, therefore, the more we see the reflection of our own faces .... Even in the most up-to-date physical concepts the anthropomorphic burden is very large. Particles still attract or repel one another, rather as do people; they 'experience' forces, are captured or escape. They live and decay. Circuits 'reject' some signals and 'accept' others; and so forth."

This is indeed far from the conventional idea that the scientist is impersonal and keeps a completely open mind. As would be expected, scientific investigators of the brain are more anthropomorphic even than nuclear physicists. We inevitably are obsessed by ideas of purpose and design.

Personal Experiences

It seems to me that discussions on scientific method often suffer from the circumstance that they are conducted as it were *in vacuo* without an intimate relationship with actual scientific problems. Since my scientific life owes so much to my conversion in 1945, if I may call it so, to POPPER'S teachings on the conduct of scientific investigations, it seems appropriate to make special reference to my personal experiences in research. Like most young scientists I started off doing research in a rule-of-thumb manner, being quite innocent of sophisticated ideas on the methodology of science, though I had accepted without qualms the inductive nature of scientific method. I was fortunate enough to be at Oxford in what we may now call the classical Sherringtonian period of research on spinal reflexes.

And I discovered that at first all could go well so long as I was lucky in my choice of problems; and this was not difficult so long as I was pursuing the general direction of the classical investigations. But later, when I ventured into new fields, I eventually began to recognize that I was making serious scientific errors, and I suffered as a consequence all the pangs that accompany the recognition of such misfortunes.

Until 1945 I held the following conventional ideas about scientific research—first, that hypotheses grow out of the careful and methodical collection of experimental data. This is the inductive idea of science deriving from BACON and MILL. Most scientists and philosophers still believe that this is the scientific method. Second, that the excellence of a scientist is judged by the reliability of his developed hypotheses, which, no doubt, would need elaboration as more data accumulate, but which, it is hoped, will stand as a firm and secure foundation for further conceptual development. A scientist prefers to talk about the experimental data and to regard the hypothesis just as a kind of working construct. Finally, and this is the important point: it is in the highest degree regrettable and a sign of failure if a scientist espouses an hypothesis which is falsified by new data so that it has to be scrapped altogether.

That was my trouble. I had long espoused an hypothesis which I came to realize was likely to have to be scrapped, and I was extremely depressed about it. I had been involved in a controversy about synapses (cf. Chapter II), believing in those days that the synaptic transmission between nerve cells was largely electrical. I admitted that there was a late, slow chemical component, but I believed that the fast transmission across the synapse was electrical.

At that time I learned from POPPER that it was not scientifically disgraceful to have one's hypothesis falsified. That was the best news I had had for a long time. I was persuaded by POPPER, in fact, to formulate my electrical hypotheses of excitatory and inhibitory synaptic transmission so precisely and rigorously that they invited falsification—and, in fact, that is what happened to them a few years later, very largely by my colleagues and myself, when in 1951 we started to do intracellular recording from motoneurones. Thanks to my tutelage by POPPER, I was able to accept joyfully this death of the brain-child which I had nurtured for nearly two decades and was immediately able to contribute as much as I could to the chemical transmission story which was the Dale and Loewi brain-child.

I had experienced at last the great liberating power of POPPER's teachings on scientific method. The remarkable result of this complete rejection of my long cherished hypothesis was that I had achieved a notorious reputation which surprisingly enough was not at all unfavourable. The story has been told on many occasions when introducing me in some dynamic pattern of operation (cf. Chapter II). Furthermore, it was not realized that, if the total information from the sensory receptors of a limb were to be fed into one particular zone of the cerebellar cortex, the integrational functioning of the neuronal machinery would be "jammed" by the overwhelming confusion of input. There must be some regional zoning of various subsets of the input so that the same input participates in many different integrational assemblages. However, before the experimental investigation this was not predicted, though when discovered it appeared so necessary and so obvious. My experience is that often I have only a very vague horizon-of-expectations (to use POPPER'S felicitous phrase) at the beginning of a particular investigation, but of course sufficient to guide me in the design of experiments; and then I am able to maintain flexibility in concepts which are developed in the light of the observations, but always of a more general character so that the horizon-of-expectations is greatly advanced and developed and itself gives rise to much more rigorous and searching experimental testing than could have been designed at the outset.

Illustrations from Neurobiological Investigations

A key point of POPPER's teaching with respect to the most effective procedures in scientific research is that it is of the utmost importance to carry out research related to an horizon-of-expectations that derives from a well developed theory extending far into the unknown. I can illustrate the manner in which this procedure has guided and aided me by giving two examples of my experiences in the recent decade or so.

The first example is the hypothesis that in the central nervous system a nerve cell can have only one kind of action at all of its synapses, for example that it can be an excitatory nerve cell exerting a postsynaptic excitatory action or it can be an inhibitory nerve cell. It cannot be ambivalent, having an excitatory action at some of its synapses and an inhibitory action at others, i.e. nerve cells can be quite sharply divided into two groups, the excitatory and the inhibitory (cf. Chapter II, Figs. 6, 7, 8). This hypothesis was put up on the basis of a very few species of these two types of nerve cells, but now in the mammalian nervous system there are over 30 species of purely inhibitory neurones and no examples of ambivalent neurones. The full story of this hypothesis and of its testing is told in my Sherrington Lectures (ECCLES, 1969b). It has stimulated much experimentation and severe attack, and there have been several claims of its refutation that later have been found to be untenable. So to date this hypothesis has not been falsified and in itself has led to the conceptual development of most interesting problems in neurogenesis that will be referred to later.

problem of the formate anomaly. The formulation of this postulate relating a critical pore size to anion permeability has stimulated much research on the ionic mechanisms of inhibitory action (cf. ECCLES, 1964, 1966d).

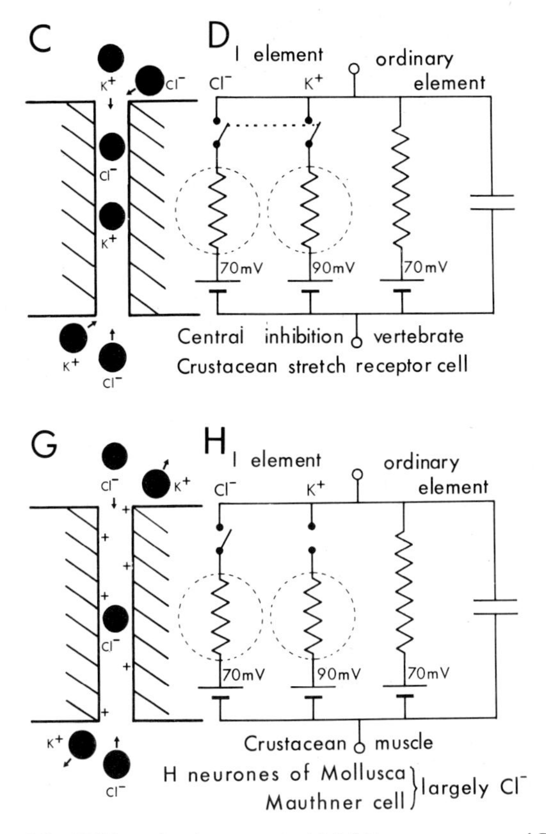

that is postulated for IPSP production at central inhibitory synapses, and D is a diagram showing that the inhibitory element is composed of potassium and chloride ion conductances in parallel, each with batteries given by their equilibrium potentials, and is operated by a ganged switch, closure of which symbolizes activation of the inhibitory subsynaptic membrane. E, F and G, H represent the conditions occurring at inhibitory synapses where there is predominantly potassium or chloride ionic conductance as indicated. It is assumed that the pores are restricted to cation or anion permeability by the fixed charges on their walls as shown (ECCLES, 1964, 1966d)

There is a further problem, namely cation permeability produced by the inhibitory synaptic transmitter, that is as yet not satisfactorily tested except in a few types of inhibitory synapses in invertebrates. However, there is the more general postulate that not only pore size, but also the fixed charges on the pores determine the ionic permeability (Fig. 26).

With a fixed positive charge (G, H), anion permeability is enhanced and cation permeability depressed, and conversely (E, F) with a fixed negative charge (cf. ECCLES, 1964). This more comprehensive hypothesis has not yet been falsified, but should be subjected to more critical testing than has yet been possible. It provides a satisfactory explanation of the experimental findings (cf. Fig. 26D, F, G) that the inhibitory ionic currents are largely carried by cations at some types of inhibitory synapses, by anions at other types, and by both anions and cations at yet others. It appears on present evidence that the critical size for hydrated ions is the same for cations and anions, but much more testing is needed.

This ionic permeability to anions and/or cations below a critical size is characteristic of the postsynaptic inhibitory synapses, and serves to differentiate them sharply from excitatory synapses. Evidently in mammalian neurogenesis there is some profound biological principle that "forbids" the synapses made by any one cell to be of two classes, its synaptic transmitter being able to open ionic gates characteristic both of excitatory synapses ($Na^+ + K^+$ permeable) and of inhibitory synapses (K^+ and/or Cl^- permeable). The conclusion to be drawn from these various examples is that there is no known exception to two general principles which can be formulated in relation to chemically transmitting synapses in the mammalian brain.

The first principle is that, at all of the synaptic terminals of a nerve cell, there is always the liberation of the same transmitter substance or substances. Strictly speaking this is DALE'S Principle, and it derives from the metabolic unity of the cell.

The second principle is that, at all of the synaptic terminals of a nerve cell, the transmitter substance opens just one type of ionic gate, that characterizing either excitatory or inhibitory synapses. Otherwise stated, a nerve cell cannot be ambivalent with respect to the essential mechanism of its synaptic action on the subsynaptic membrane.

In contradistinction to the first principle, there is no simple explanation for the rigid operation of the second principle. For example, it cannot simply be derived from the first principle, because a transmitter substance such as acetylcholine acts as an excitatory type of transmitter substance at some synapses (sympathetic ganglia, neuromuscular junctions and the synapses of motor axon collaterals on the Renshaw cells in the spinal cord, and as an inhibitory type of transmitter at others (vagus on heart, synapses on H- and D-cells in Mollusca). It seems that there must be some principle of neurogenesis whereby the outgrowing axonal branches of a neurone can make effective synaptic contacts only of an excitatory or of an inhibitory type.

Firstly, it is postulated that already at that very early stage of neural outgrowth the neurone is specified as excitatory (E) or inhibitory (I)

(ECCLES, 1970b). It is possible then to formulate a postulate about the way in which such a specified neurone is constrained to establish synaptic contacts exclusively of the one or the other type (ECCLES, 1969b).

The most probable postulate (Fig. 27A) seems to be that, by chemical sensing, the growth cones of an I-neurone search out patches on the surface of a neurone that are already specified as being inhibitory, there being the consequent development of a functional inhibitory synapse—and similarly for the growth cones of E-neurones and the formation of

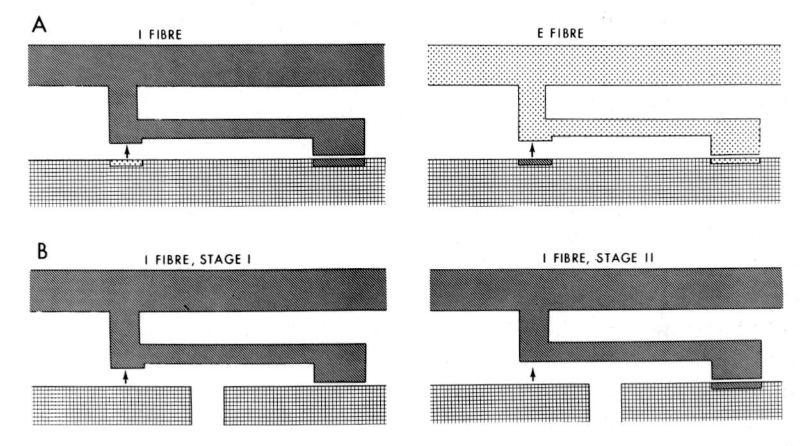

Fig. 27A and B. Diagram of two possible developmental procedures by which a nerve cell establishes either purely inhibitory or purely excitatory synaptic connections. Inhibitory cells are shown with dark toning, excitatory with light. In A the axonal outgrowth from an already specified inhibitory cell to the left rejects a contact with an already specified excitatory synaptic patch and grows on to find an acceptable inhibitory synaptic patch. The reciprocal situation is seen in A to the right. B illustrates an alternative postulate, but only for an already specified inhibitory cell, as described in the text

functional E-synapses. Certainly the surface of a muscle fiber is covered by preformed cholinoceptive patches before the motor nerve fibers establish synaptic contacts.

An alternative postulate (Fig. 27B) is that the preformed I-growth cones sense out appropriate neurone surfaces that are not yet specified with patches, but are identified by some other chemical criterion, in the way that is observed with growth of nerve fibers in tissue culture. The I-growth cones then make effective I-synapses by creating in the subsynaptic membrane of these appropriate neurones the receptor patches for the I-transmitter substance and the associated ionic gates that give the fully functional inhibitory synapses; and similarly with the E-growth cones. We have to make the additional postulate that the presynaptic terminal of an I-cell together with the I-transmitter liberated therefrom can create not only receptor patches for the I-transmitter, but also the ionic gates of inhibitory character, i.e. for K^+ and/or Cl^- and not for Na^+; and similarly for the E-synapses—Na^+ plus K^+ and not Cl^-.

In general these postulates give essentially the same result with respect to inhibitory interneurones—namely that the I-cell cannot form effective E-synapses, since it has the wrong transmitter to open the E-gates on these particular synaptic surfaces. These postulates are as yet purely speculative, but they have the merit of defining problems susceptible to scientific experiment and in providing new insights in relation to observations that have already been reported.

These examples from my own experiences in neurophysiology illustrate the manner in which I have endeavoured to follow POPPER in the formulation and in the investigation of fundamental problems in neurobiology. I hope that they illustrate the sense of liberation and adventure that I have derived from this. Furthermore, I think they have enabled me to progress much further and faster in my efforts to understand some operative features of the central nervous system.

Scientific Diseases

I am aware that it can be regarded as unseemly for me even to mention scientific diseases; nevertheless I do so because I fear that they present a most serious and insidious threat to science. I believe that the diseases arise from two main sources: the failure to understand the nature of the scientific method; the failure of scientists to appreciate that science is a shared enterprise and adventure of mankind, and not a means to achieve one's personal advantage and fame. I admit that this list is based on personal predilection, and I welcome any thoughtful criticism that it may evoke, though of course it will also provoke emotional outbursts!

There is an erroneous belief that status as a scientist is given by the cost and elaboration of equipment. As a consequence there can be a competition to spend beyond all reason, and a display of equipment for its own sake and not in relation to the requirements of the scientific investigation. One is reminded of the ostentatious display of social climbers!

The enslavement to equipment brings in its train the disease in which the experiments actually being performed are chosen by the equipment, and not by the investigator. Scientific papers become reports of the usage of instruments, not of the efforts to understand some aspect of nature. In its most aggravated form the most elaborate equipment is used in the mistaken belief that it is thereby possible to retrieve information that is already lost by poor experimental procedures.

The exploitation of science for personal ascendency gives rise to antagonism and even to animosity with respect to rival investigators who are regarded as threats to the cherished ascendency, and not as colleagues in an enterprise. It also gives rise to unseemly quarrels about priority of discovery. When fellow scientists attempt to falsify your hypothesis, it must be regarded even as a compliment that they regard your hypothesis as worth the efforts of criticism and not something just to be ignored.

Arrogance is one of the worst diseases of scientists and it gives rise to statements of authority and finality which are expressed usually in fields that are completely beyond the scientific competence of the dogmatist. It is important to realize that dogmatism has now become a disease of scientists rather than of theologians. POPPER would remind us that we have to be humble and recognize the limitations of our most penetrating efforts to understand nature; and we must never claim to have given a definitive explanation but only to have stated an hypothesis that conforms with all existing knowledge. The best that we can do experimentally is to corroborate our hypothesis, not to confirm it. It is relevant to quote from POPPER (1962):

> "But although the world of appearances is indeed a world of mere shadows on the walls of our cave, we all constantly reach out beyond it; and although, as DEMOCRITUS said, the truth is hidden in the deep, we can probe into the deep. There is no criterion of truth at our disposal, and this fact supports pessimism. But we do possess criteria which, *if we are lucky*, may allow us to recognize error and falsity. Clarity and distinctness are not criteria of truth, but such things as obscurity or confusion *may* indicate error. Similarly coherence cannot establish truth, but incoherence and inconsistency do establish falsehood. And, when they are recognized, our own errors provide the dim red lights which help us in groping our way out of the darkness of our cave."

It has been well said that truth comes out of error much more readily than out of confusion.

That scientific status devolves from the skill in carrying out the technical procedures in experiments rather than from the imaginative and intellectual activities that are involved in the discovery and clarification of problems, in the designing of experimental tests, and finally in the clear presentation of the resulting problem situation in a scientific paper. Such an extraordinary misunderstanding of course derives from ignorance of the nature of science, and from confusing science with technology. It leads to much display of techniques that are not significantly applied in science. This disease afflicts in particular the immature who have not yet realized what science is about. However, many scientists never mature, but go on suffering from this children's disease.

Another disease is the materialistic or mechanistic philosophy held by so many scientists, who reject or ignore all the phenomena of conscious experience, and thus display a complete misunderstanding of the working of the brain, which all must agree is essentially concerned in science! I am reminded of some physical scientists who facetiously regard life as but a disease of matter and would restrict the scope of science to the inorganic world! Scientists who reject the phenomena of conscious experience from the domain of science suffer from a comparable mental scotoma, for they reject for example all such perceptual experiences as light, color, sound, touch, pain, and all the combinations of these and their recall in memory that give for example the perceptual data of the experiences that I for example have. They would also reject the imagination, the critical thought, the evaluation, and the judgement which are so vitally concerned in conscious activities as one wrestles with scientific problems and eventually creates some scientific paper that reports the experimental testing of some hypothesis with the corroboration or falsification deriving therefrom.

It is not of course implied that I am free from these diseases. I am merely describing diseases as a clinician who may himself be sick. Nor am I suggesting that scientists afflicted by such diseases are thereby rendered ineffective, but only that, as with physical disease, they are handicapped to some degree. However, in the worst cases the disease may prove fatal to the scientific life of the victim.

It will be evident that these diseases arise from a misunderstanding of the nature of science. The treatment that I would suggest is quite simple—namely to read and meditate upon POPPER's writings on the philosophy of science and to adopt them as the basis of operation of one's scientific life! I am not of course advocating any slavish adherence to every detail, but rather the enlightenment of general principles. POPPER himself has summed up his view of the method of science in three words: problems—theories—criticism.

POPPER puts up extremely rigorous principles of operation in respect of the testing of scientific hypotheses, but he mitigates their severity by providing encouragement even when there are repeated failures in the attempted solution of a problem (POPPER, 1963).

> "Thus we become acquainted with a problem only when we have many times tried in vain to solve it. And after a long series of failures—of producing tentative solutions which turn out not to be acceptable solutions of the problem—we may even have become experts in this particular problem. We shall have become experts in the sense that, whenever somebody else offers a new solution—for example, a new theory—it will be either one of those theories which we have tried out in vain (so that we shall be able to explain why it does not work) or it will be a new solution, in which case we may be able to find out quickly whether or not it gets over at least those standard difficulties which we know so well from our unsuccessful attempts to get over them.
>
> My point is that even if we persistently fail in solving our problem, we shall have learned a great deal by having wrestled with it. The more we try, the more we learn about it—even if we fail every time. It is clear that, having become in this way utterly familiar with a problem—that is, with its difficulties—we may have a better chance to solve it than somebody who does not even understand the difficulties. But it is all a chance: in order to solve a difficult problem one needs not only some understanding but also some luck."

General Summary

As a general summary I can quote from an after-dinner talk that I gave some two years ago (ECCLES, 1966a). It is a verbatim transcript of a tape recording, hence I hope I may be excused for its rather racy style:

"What are the advantages of the Popperian method of giving as precise a formulation of the hypothesis as possible—going far beyond the data? You are to put up some imaginative construct far beyond any evidence, but based, of course, securely upon what there is—not falsified by anything that exists, but going far beyond anything that is yet known.

First of all, it challenges falsification. That helps a lot. Your scientific colleagues say: 'I don't believe that. I'm going to do some further experiments with this.' Well, that's helping science, and the thing becomes immediately scientifically interesting because you've been challenged.

Secondly, it greatly economizes the experimental effort by giving it significant direction. You see how to design your experiments so as to test ideas that you have already formulated. You shouldn't fiddle around hoping something will turn up. Nothing does as a rule. All you'll get is a large number of publications which will be lost quite quickly because they haven't been built into clear stories. The things that have survived from the past, that exist as classical works, are those linked to clear and significant stories or hypotheses that have been meaningful.

Of course, if your hypothesis is falsified or has to be remodelled that's fine, too, because even in the denouement science has been well served. You've learned something. By putting up a clear story, doing the experiment, replacing it by another clear story, you are progressing. And that is the essential method of science; it's the deductive, critical method of POPPER—the hypothetical-deductive method. You hypothesize upon the basis of what you know, but with implications far beyond that, and then you proceed to criticize and to test them experimentally as rigorously as you can. Don't bolster up your hypotheses; try to kill them. Attack them where they are weakest, don't wait for somebody else to do it. It's much better to bring down your own hypothesis, and you can do so better than the other chap. Meanwhile, your progress has been conceptually and experimentally economical because your experiments have been designed around ideas and stories.

In the end—let's be clear about this—you must realize that the important thing is to recover from erroneous scientific beliefs, and this you do if you realize that these hypotheses have been extremely valuable scientifically, although falsified. They have contributed positively. In fact the only thing you can do for sure in science is that you can disprove something. You can't prove things. You can test them. But a complete and final proof of anything except triviality is impossible in science. So what we need firstly is imagination to go beyond the data and put up hypotheses which are built securely upon what we know and which lead on into a future which organizes testing experiments."

Finally I would like to conclude with a most wise and authoritative statement by POPPER (1962) that summarizes so remarkably the approach we should have in our attempts in "the understanding of nature."

"What we should do, I suggest, is to admit that all knowledge is human; that it is mixed with our errors, our prejudices, our dreams, and our hopes; that all we can do is to grope for truth even though it be beyond our reach. We may admit that our groping is often inspired, but we must be on our guard against the belief, however deeply felt, that our inspiration carries any authority, divine or otherwise. If we thus admit that there is no authority beyond the reach of criticism to be found within the whole province of our knowledge, however far it may have penetrated into the unknown, then we can retain, without danger, the idea that truth is beyond human authority. And we must retain it. For without this idea there can be no objective standards of inquiry; no criticism of our conjectures; no groping for the unknown; no quest for knowledge."

Chapter VIII

Man, Freedom and Creativity

1. Free-Will

The self has other properties besides purely being able to experience things as described in Chapters IV and V. It can also do things. And so we are confronted with the problem of free-will. We can argue: do we have free-will, or are all our decisions the manifestations of something that is in-born, plus all our training or conditioning? If a statement is made: "I do not have free-will," two inferences can be made: this statement could be a result of a conditioning, the subject being well-trained to repeat it like a parrot, which makes it devoid of significance; or the subject may claim that he has free-will just to make this statement, which is an arbitrary and self-stultifying statement.

Whatever their expressed beliefs, people in fact behave as if they have free-will, which, I think, is best shown by trivial things in life—whether, for example, I can take a coin and put it here or there, turn it over or do the most capricious things with it. We demand of our nervous and muscular equipment that we can at will cause it to carry out any desired actions which are within our power, no matter how trivial or capricious. Remarkable observations can be made by evoking movements by stimulating the motor areas of the brain in a conscious subject. Though these movements were elicited as normally by messages from the motor cortex, the subject distinguishes quite clearly between them and those that he voluntarily initiated. He will say, "This movement is due to something done to me and not something done by me."

There is no answer to such a question as: Is any reconciliation possible between the direct experience that an act of will can call forth a muscular movement, and on the other hand, the scientific account whereby such a muscular movement results from an activity of nerve cells in the brain, which in turn is relayed by nerve impulses eventually to muscles? It is my contention that these questions concerning the problem of brain-mind liaison have been wrongly posed.

I have a direct experience that my thought can lead to action. I can decide on a particular action, perhaps of the most trivial nature, and my muscular movements can be directed towards accomplishing that act. I have no experience of the manner in which my willing leads to action. Naturally, scientific investigation can be applied to study the sequence of events leading to movement: the discharges of pyramidal neurones of the motor cortex; the propagation of these impulses down the pyramidal tract; the discharge of impulses so evoked from motoneurones and propagation out to muscles leading to the muscular activation and eventual contraction. There would be no evidence supporting my belief that my body does carry out my willed movements. However there is an evident inadequacy in this documentation. As yet we have no knowledge about the neuronal pathways upstream from the pyramidal cells of the motor cortex (PHILLIPS, 1966). Presumably there are complex patterns of neuronal activity converging onto the motor cortex, and these would be of immense complexity as followed upstream.

A remarkable example of extremely fine voluntary control has been described by BASMAJIAN (1963). The rhythmic discharges of a single motoneurone can be visualized by a subject, being led off by an electrode in one of his muscles and displayed after amplification on a cathode ray screen and by a loudspeaker. After a training period of some hours he is able to vary the frequency of discharge of that single cell at will and to cause it to discharge in preference to other nerve cells to that muscle, and he even can do this in the absence of the visual or auditory aids. These experiments provide remarkable examples of the skill and finesse of the voluntary control of movement.

Curiously enough, the most compelling evidence for the belief in free-will comes when there is some failure in the control of movement. If I find that I cannot direct my muscular movements in some willed manner, as, for example, taking a coin and putting it on some particular square of a chess-board, I would immediately recognize this as due to some disorder in my nervous system, which is called compulsion neurosis. I would consult a neurologist or a psychiatrist; and this would be the reaction of all normal human beings in a civilized society. Thus the belief that it is possible to exert a conscious control of movement is best demonstrated by the response to any untoward limitation of this so-called "freedom of the will."

It is not contended that all action is willed. There is no doubt that a great part of the skilled activity devolving from the cerebral cortex is stereotyped and automatic, as for example the routine driving of a car, and may be likened to the control of breathing by the respiratory centers. But it is contended that it is possible voluntarily to assume control of such actions, even of the most trivial kind, just as we may within limits exercise a voluntary control over our breathing.

The principal grounds for the theoretical belief that this control is an illusion are derived from the assumptions that both physics and neurophysiology give a deterministic explanation of all events in the brain and that we are entirely within this deterministic scheme. In this context reference may be made to the discussion by POPPER (1950) in which he concludes that not only quantum physics but even "classical mechanics is not deterministic, but must admit the existence of unpredictable events." A similar argument has been developed by MACKAY (1966). The neurophysiology of a deterministic character is merely a primitive reflexology, and not related at all to the dynamic properties of the immense neuronal complexities of the brain. There are thus no sound scientific grounds for denying the freedom of the will, which, ironically, must be assumed if we are to act as scientific investigators.

One can surmise from the extreme complexity and refinement of its organization that there must be an unimagined richness of properties in the active cerebral cortex. Meanwhile, I continue to believe in the freedom of my will, though its mode of operation cannot at present be explained scientifically; nevertheless it is important to speculate in order to develop ideas about possible or conceivable physiological mechanisms.

The Neurophysiological Problem of Will

An important neurophysiological problem arises as soon as we attempt to consider in detail the events that would occur in the cerebral cortex when, by exercise of "will," some change is induced in the response to a given situation. As argued above, in a situation where "will" is operative, there will be a changed pattern of discharge down the pyramidal tract and this change must be brought about because there is a change in the spatio-temporal pattern of influences playing upon the pyramidal cells in the motor cortex. If the "will" really can modify our reactions in a given situation, we have somewhere in the complex patterned behaviour of the cortex to find that the spatiotemporal pattern which is evolving in that given situation is modified or deflected into some different pattern.

Quantitative Aspect of Spread of Activity in Neuronal Networks

In the formulation of problems concerning activity in neuronal networks it is of value to have a model of the simplest possible network (Fig. 28 A; cf. BURNS, 1951, Fig. 10). Each neurone is assumed to have only two excitatory synaptic knobs on its surface and its axon has only two excitatory knobs on two other neurones. Further it is assumed that the synaptic connexions so formed are of a two-dimensional pattern that allows the neurones to be arranged schematically in the rectangular net-like form of Fig. 28A, where it will be noticed that there is virtual radial symmetry from any point and the possibility of indefinite extension in every direction. If each neurone receives and gives three synaptic

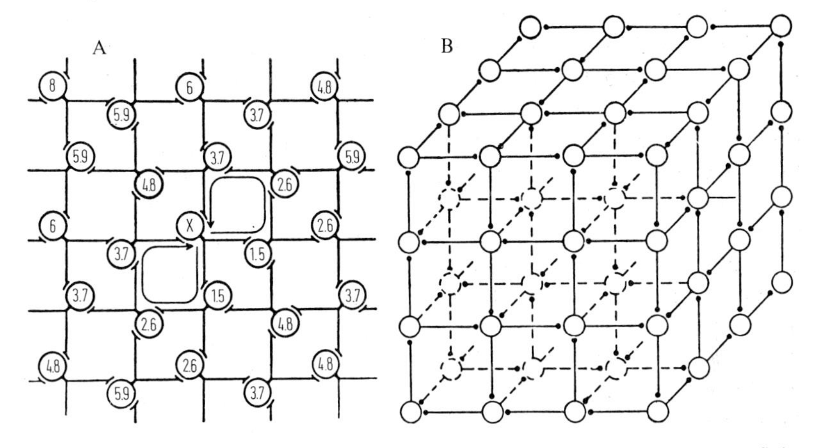

Fig. 28. A Schematic model of simplest type of neuronal network that would give indefinite outward spread from an excited focus and also provide closed self-re-exciting chains of all degrees of complexity. Each neurone is assumed to have only two synaptic knobs on its surface and in turn to have an axon which has only two knobs on other neurones. This net could be extended in any direction and there would be virtual radial symmetry from any point. For example, the numerals on each neurone give the number of synapses traversed in leading to its first (and second) activation in spread from neurone X, and the two simplest closed chains (4-neurone arcs) are shown by arrows. Next simplest are six 8-neurone arcs, then 12-neurone arcs, and so on. B Diagram with similar conventions to those of A but with each neurone receiving and giving three synaptic contacts. Unfortunately, owing to the complexity of the diagram, it has been impossible to draw the full connections for more than a very few neurones. Moreover, only the surface neurones of the deeper layers (3 and 4) are shown. Connections and cells in the depth of the cube are shown by broken lines. Note same convention as in A, namely that in any plane the adjacent transmission lines alternate in direction (Eccles, 1953)

contacts, a similar construction with radial symmetry is possible in a three-dimensional network (Fig. 28B). As shown in Fig. 28 these constructions would give alternating direction of transmission in the successive lines in any plane. Similarly, if each neurone gave and received n synapses, the pattern could be accommodated to an n-dimensional network.

The problem of the mode of action of the "will" can be simplified and sharpened by considering firstly the behaviour of a single neurone in the active neuronal network of the cortex. Suppose some small "influence" were exerted at a node that would make a neurone discharge an impulse at a level of synaptic excitation which would otherwise have been just ineffective, that is, in general to raise the probability of its discharge. Such a discharged impulse would in turn have an excitatory effect on all the other nodes on which it impinges, raising the probability of their discharge, and so on. If we assume, as above, that the transmission time from node to node occupies 1 msec, then, even on the two-dimensional net of Fig. 28 A, a spread to a large number of neurones is possible in, say, 20 msec, a time that is chosen because it is at the lower limit of duration of discrete mental events.

In order to frame a precise problem, we can firstly consider the schematic neuronal networks of Fig. 28 which are assumed to be cortical neurones—both the pyramidal cells and the very numerous stellate cells (cf. Figs. 2, 3 A, 32 A). We make the postulates that at zero time a neurone (for example X in Fig. 28 A) is caused to discharge an impulse into the quiescent network and that activation of one synapse is adequate to cause any neurone to discharge an impulse. For the network of Fig. 28 A, the total number of neurones, N, caused to discharge impulses is given by the formula (SAWYER, 1951):

$$N=2m^2-2m+2$$

where m is the number of nodes traversed. In 20 msec $m=20$, the internodal time being assumed as 1 msec; hence the number of neurones activated is 762.

On the same assumptions, but with a multi-dimensional network constructed according to the conventions of Fig. 28, the number of activated neurones, N, is given, where m is large relative to n, by the general formula (SAWYER, 1951):

$$N\sim\frac{2n}{n!}\,m^n$$

when $m=20$ (i.e. within 20 msec) and with $n=3$ (Fig. 28 B), N is of the order 10^4. With $n=4$ and 5 respectively, N is of the order 10^5 and 8×10^5.

These calculations are intended merely to give some indication of the large number of cortical neurones that could be affected by a discharge originating in any one. In order to apply them to our problem of how "will" *could* act on the cerebral cortex, it is necessary to take into account the evidence that "will" can act on the cortical neuronal network only when a considerable part of it is at a relatively high level of excitation, i.e. we have to assume that, for "will" to be operative, large populations of cortical neurones are subjected to strong synaptic bombardments and are stimulated thereby to discharge impulses which bombard other neurones (cf. Figs. 10, 12). Under such dynamic conditions it may be conservatively estimated that, out of the hundred or more synaptic contacts made by any one neurone, at least four or five would be *critically effective* (when summed with synaptic bombardments by other neurones) in evoking the discharge of neurones next in series. The remainder would be ineffective because the recipient neurones would not be poised at this critical level of excitability, being either at a too low level of excitation, or at a too high level, so that the neuronal discharge occurs regardless of this additional synaptic bombardment. Thus at any instant the postulated action of the "will" on any one neurone would be effectively detected by the "critically poised neurones" on which it acts synaptically.

So long as the assumed number of *critically effective* synaptic excitatory actions by each neurone is kept at the low levels used in the above calculations, it is probable that the conventions of the network structures of Fig. 28 give an approximate method of allowing for all the mass of feed-back connections that occur in the closed-chain linkages of the cerebral cortex (LORENTE DE NÓ, 1933, 1934, 1943). Further, since the cortex is approximately 3 mm thick and the mean density of neurones 40,000 per sq. mm of surface (THOMPSON, 1899), the spread to some hundreds of thousands of neurones can be treated as spreading indefinitely in all directions without serious restriction by the sheet-like structure of the cortex. Hence we may conclude that, when a region of the cortical neuronal network is at a high level of activity, the discharge of an impulse by any one neurone will have contributed directly and indirectly to the excitation of hundreds of thousands of other neurones within the very brief time of 20 msec.

A Neurophysiological Hypothesis of Will

As a restatement of the conclusion of the preceding section we may say that in the active cerebral cortex within 20 msec the pattern of discharge of even hundreds of thousands of neurones would be modified as a result of an "influence" that initially caused the discharge of merely one neurone. But further, if we assume that this "influence" is exerted not only at one node of the active network, but also over the whole field of nodes in some sort of spatio-temporal patterning, then it will be evident that potentially the network is capable of integrating the whole aggregate of "influences" to bring about some modification of its patterned activity, that otherwise would be determined by the pattern of afferent input and its own inherent structural and functional properties. Such integration would occur over hundreds of thousands of nodes in a few milliseconds, the effects exerted on any and every node being correlated in the resultant patterned activity of the surrounding hundreds of thousands of neurones. Thus in general, the spatio-temporal pattern of activity would be determined not only by (i) the micro-structure of the neural net and its functional properties as built up by genetic and conditioning factors and (ii) the afferent input over the period of short-term memory, but also (iii) by the postulated "field of mind influence." For example, in Fig. 29 the spatio-temporal pattern determined by factors (i) and (ii) is shown diagrammatically by the shaded structure bounded by the continuous line, while a possible modification by factor (iii) is indicated by the paths outlined by broken lines at B and C. Fig. 29 can be considered as showing boundaries of multilane neuronal traffic as indicated in Figs. 10 and 12.

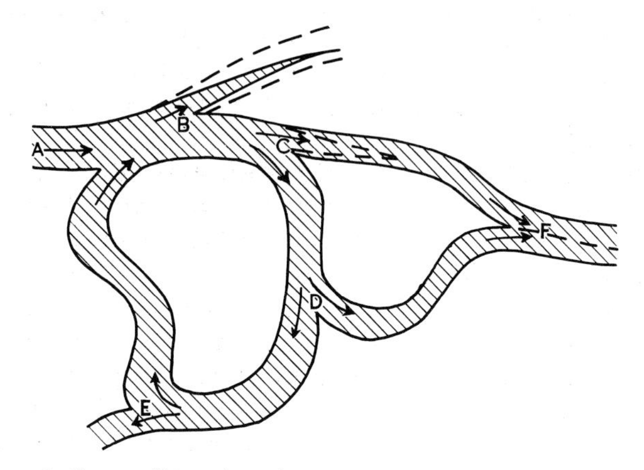

Fig. 29. Schematic diagram illustrating a fragment of a spatiotemporal pattern of neuronal activity in the cerebral cortex, and drawn in outline according to the same conventions as those adoped in Figs. 10 and 12. It is assumed that by will the pattern can be altered as shown by the broken lines. Full description in text. (ECCLES, 1953)

It can be claimed that no physical instrument would bear comparison with the postulated performance of the active cerebral cortex as a detector of minute "fields of influence" spread over a microscopic pattern and with temporal sequences of milliseconds. The integration, within a few milliseconds, of "influences" picked up at hundreds of thousands of nodes would be unique, particularly when it is remembered that the integration is no mere addition, but is exerted to modify in some specific way "a shifting harmony of sub-patterns" of neuronal activity, achieving expression through the modifications so produced.

Thus, the neurophysiological hypothesis is that the "will" modifies the spatio-temporal activity of the neuronal network by exerting spatio-temporal "fields of influence" that become effective through this unique detector function of the active cerebral cortex. It will be noted that this hypothesis assumes that the "will" or "mind influence" has itself some spatio-temporal patterned character in order to allow it this operative effectiveness.

The Physical Implications of the Hypothesis

When considering the manner in which mind could operate on matter, EDDINGTON (1939) discussed two hypotheses.

(i) It was postulated that mind could control the behaviour of matter within the limits imposed by HEISENBERG'S Principle of Uncertainty (cf. EDDINGTON, 1935). EDDINGTON rejected this partly because the permitted range would be exceedingly small. Presumably he was thinking of an object as large as a neurone. However, a neurophysiologist would now consider the much smaller synaptic vesicle (cf. Figs. 3, 4, 31) as the key structure on which a "mind influence" might work. The synaptic vesicle is approximately a sphere 400 Å in diameter and so would have a mass of about 3×10^{-17} g. If, as EDDINGTON implies, the uncertainty principle is applicable to an object of this size, then it may be calculated that there is an uncertainty in the position of such an object of about 50 Å in 1 millisecond. These values are of interest since 50 Å is approximately the thickness of the presynaptic membrane across which the vesicle discharges its content of specific transmitter substance (cf. Fig. 4D).

Furthermore, as shown above, minute "influences" thus exerted on a large population of neurones would be rapidly integrated in the form of a changed spatio-temporal pattern of activity in the neuronal net. There is thus in the active cortex a mechanism that could enormously amplify minute effects exerted on the individual synaptic vesicles, provided of course, as postulated above, these influences have some "meaningful" pattern and are not random. It is therefore possible that the permitted range of behaviour of a synaptic vesicle may be adequate to allow for the effective operation of the postulated "mind influences" on the active cerebral cortex. However, EDDINGTON rejected this hypothesis for the further reason that it involved a fundamental inconsistency. First, behaviour according to chance was postulated in making a calculation of the permitted limits according to the uncertainty principle, then it was restricted or controlled by a non-chance or volitional action (the mind influence), which necessarily must be introduced if mind is to be able to take advantage of the latitude allowed by the uncertainty.

(ii) As a consequence of this rejection, EDDINGTON was led to an alternative hypothesis of a correlated behaviour of the individual particles of matter, which he assumed to occur for matter in liaison with mind. The behaviour of such matter would stand in sharp contrast to the uncorrelated or random behaviour of particles that is postulated in physics, and, as he stated, may be "regarded by us as something 'outside physics'" (EDDINGTON, 1939).

Either of EDDINGTON'S hypotheses could serve as the physical basis of the neurophysiological hypothesis that has here been developed for mind-brain liaison. This latter hypothesis of mind-brain liaison has the merits of relating the occasions when the mind can operate on the brain to the observed high level of neuronal activity during consciousness, and of showing how an effective action could be secured by a spatio-temporal pattern of minute "influences." If the neuronal activity of the cerebral cortex is at too low a level, then liaison between mind and brain ceases. The subject is unconscious as in sleep, anaesthesia, coma. Perception and willed action are no longer possible. Furthermore, if a large part of the cerebral cortex is in the state of the rigorous driven activity of a convulsive seizure, there is a similar failure of brain-mind liaison, which is likewise explicable by the deficiency of the sensitive detectors, the critically poised neurones.

General Discussion of Hypothesis of Free-Will

It will be sufficiently evident that the hypotheses here developed are of a fragmentary and tentative character, but it is hoped that they may be of value in further theoretical developments on mind-brain liaison. An outstanding problem for consideration would concern the postulated action of the mind in a spatio-temporal pattern, for presumably it must so act if it is to cause significant modification in patterned activity of the cortex. However, that problem would appear less formidable if there were a sufficiently rapid and detailed feed-back from brain action to mind, which in any case must be assumed for perception.

It will be objected that the essence of the hypothesis is that mind produces changes in the matter-energy system of the brain and hence must be itself in that system (cf. SCHRÖDINGER, 1951). But this deduction is merely based on the present hypotheses of physics. Since these postulated "mind influences" have not been detected by any existing physical instrument, they have necessarily been neglected in constructing the hypotheses of physics, as was recognized by EDDINGTON (1939). It is at least claimed that the active cerebral cortex conceivably could be a detector of such "influences" even if they existed at an intensity below that detectable by physical instruments.

The present hypotheses would offer an explanation of the high development of matter-mind traffic in the active human cerebral cortex, the development including not only continuous operation but also exquisite subtlety in transmission. Both these features would receive explanation on the basis of the interlocking, integrating, and everchanging pattern of activity formed by the numerous sensitively poised detectors (probably hundreds of millions) that exist in the cortex during states of consciousness.

It should be pointed out that, in the discussion of the functioning of the brain in Chapters II, III, it has initially been regarded as a "machine" operating according to the laws of physics and chemistry. In conscious states (Chapter IV) it has been shown that it could be in a state of extreme sensitivity as a detector of minute spatio-temporal fields of influence. The hypothesis is here developed that these spatio-temporal fields of influence are exerted by the mind on the brain in willed action. If one uses the expressive terminology of RYLE (1949), the "ghost" operates a "machine," not of ropes and pulleys, valves and pipes, but of microscopic spatio-temporal patterns of activity in the neuronal net woven by the synaptic connections of ten thousand million neurones, and even then only by operating on neurones that are momentarily poised close to a just threshold level of excitability. It would appear that it is the sort of machine a "ghost" could operate, if by ghost we mean in the first place an "agent" whose action has escaped detection even by the most delicate physical instruments.

But even if the hypotheses of brain-mind liaison here developed are on the right track, they are still extremely inadequate. For example, we have no concept of the nature of the mind that could exert these "ghost-like" influences. Again, the slight and irregular telepathic communications being excepted, it is not possible to answer the question: how is it that a given self is in liaison exclusively with a given brain? A further problem concerns the presumed spatio-temporal patterning of the mind. For example, is this altered, as may be operatively desirable, as the microstructure of the brain alters with developing experience and the consequent storage of memories?

2. Freedom and Creativity

It would be generally accepted that creative imagination is the most profound of human activities. It provides the illumination of a new insight or understanding. In science, creative imagination gives that revelation of a new hypothesis embracing and transcending the older hypotheses. There is an immediate aesthetic appeal in its simplicity and scope. Nevertheless it has to be subjected to rigorous criticism and experimental testing. In the most striking examples the illumination has had the suddenness of a flash, as with KEKULE and the benzene ring, DARWIN and the theory of evolution, HAMILTON and his equations. Yet with most of the great scientific hypotheses there was not this instantaneous and apparently miraculous birth of a "brain child." Rather were they developed in stages, being perfected and shaped by critical reason, as with PLANCK and the quantum theory and with EINSTEIN and the theory of relativity. Nor is the suddenness of illumination any guarantee of the validity of a hypothesis. I have had only one such sudden illumination

—the so-called Golgi-cell hypothesis of inhibition (Brooks and Eccles, 1947)—and some years later it was proved false (Eccles, 1953)!

If I reflect on the happenings during a scientific investigation, I find that there is incessant "traffic" between my conscious experiences and the objects and events in the external world. For example, from the framework of scientific knowledge I derive some ideas about what I should observe under certain experimental conditions. I plan these conditions and then by means of controlled movements proceed to actualize these conditions. My observations or conscious experiences of the ensuing results are correlated and evaluated in rational and critical thought against my original ideas, and further experiments are planned and executed, and so on. The consequence is that my scientific ideas or hypotheses are enriched, or changed or falsified. My scientific activity is thus seen to be essentially an affair of my rational and conceptual thought together with my exercise of willed movement and my sensory perception.

Before attempting to picture the brain activities that underlie creative imagination, it is important to realize that such illuminations, whether flash-like or with a more gradual and controlled development, come only to minds that have been prepared by the assimilation and critical evaluation of the knowledge in that particular field. One can deliberately seek to experience some new imaginative insight by pouring into one's mind hypotheses and the related experiments and then relax to give opportunity for the subconscious processes that may lead to the illumination in consciousness of a new insight. Such illuminations are often fragmentary and require conscious modification, or so erroneous as to invite immediate rejection by critical reason. Nevertheless, they all give evidence of the creativeness of the subconscious mind.

It may now be asked: what kind of activity is occurring in the brain during this creative activity of the subconscious mind, and how eventually does this creative activity flash into consciousness? Let us consider firstly the pre-requisites for such cerebral action. The wealth of stored memories and critical evaluations implies that in the neuronal network there is an enormous development of complex highly specialized engrams (cf. Chapter III) whose permanency derives from the postulated increases in synaptic efficacy. We may say that these "plastic" patterns give the know-how of the brain. We become expert in some fields of knowledge by virtue of the wealth and sublety of engrams that may extend over the greater part of the cortex. When pondering deeply on some problem in this field there must be an unimaginably complex and vivid interplay in the activated patterns. One can speculate further that some failure in the synthesis of these patterns or some conflict in their inter-relationship is the neuronal counterpart of a problem that clamours for solution.

Such are the pre-requisites leading to creative insight. We may surmise that the "subconscious operation of the mind" is dependent on the continued intense interplay of these patterns of neuronal activation. We have seen (Chapter III) that on repeated activation of any neuronal pattern there tends to be a progressive change in the basic plastic pattern or engram, particularly on account of interactions with other patterns. Thus we can expect that new emergent patterns will arise during the subconscious operation of the mind. Should an emergent pattern have an organization that combines and transcends the existent patterns we may expect some resonant-like intensification of activity in the cortex, which would bring the new transcendent pattern to conscious attention, where it would appear as a bright new idea born of creative imagination.

Then begins the process of conscious criticism and evaluation, which seeks to discover flaws in the new idea, and also to discover if it is consistent with the existing knowledge. This done, there comes the crucial stage of the design and carrying out of experiments that test predictions derived from the new idea. We may say that a creative imagination is particularly fruitful if it develops new hypotheses that are remarkable for their generality and for the manner in which they stand up to crucial experimental tests.

Finally, we may ask: what are the characteristics of a brain that exhibits remarkable power of creative imagination? In attempting an answer we are more than ever in the realm of speculation, but certain general statements can be made, though their inadequacy is all too apparent. There must, firstly, be an adequate number of neurones, and, more importantly, there should be a wealth of synaptic connection between them, so that there is, as it were, the structural basis for an immense range of patterns of activity. It is here that the inadequacy of explanation is so evident. There is but a poor correlation between brain size and intelligence, but in this assessment one is assuming a proportionality of brain size and neurone population. Furthermore, a chimpanzee brain may have a neurone population as high as 70% of a human brain, yet it displays almost no creative imagination. Secondly, there should be a particular sensitivity of the synapses to increase their function with usage (cf. Chapter III) so that memory patterns or engrams are readily formed and are enduring. Both of these properties will ensure that eventually there is built up in the brain an immense wealth of engrams of highly specific character. If added to this there is a peculiar potency for unresting activity in these engrams so that the spatio-temporal patterns are continually being woven in most complex and interacting forms, the stage is set for the deliverance of a "brain child" that is sired, as we say, by creative imagination.

3. Man and Freedom [1]

The last five hundred years have witnessed a tremendous human achievement. By the methods of scientific investigation European man has learned so many of the processes of nature that he has virtually become master of his environment. Scientific knowledge of energy and of matter has given him the control of immense resources of power and synthetic materials excellently suited for every purpose. The machine production of goods is now so enormous and varied that the manual worker has at his disposal means for comfortable living such as even the wealthiest did not possess a few generations ago. Why is it, then, that in spite of all this achievement, this undeniable improvement in the material conditions of living, this age is one of disillusionment and of a sense of frustration and unhappiness in life?

In order to answer this question we must look for the motives behind the achievements. Throughout those centuries man's vision has become progressively more "this worldly." Man has regarded himself more and more as a material being, whose life requires nothing more than the satisfaction of his material needs. The great effort of those centuries has been directed towards that end, and it cannot be denied that it has been an unparalleled success and good in itself. At the same time any belief that man has a spiritual nature as well has played progressively less part in the practical affairs of life. Religious beliefs now remain merely as personal idiosyncracies divorced from politics, economics, and the ordering of society. With this spiritual decay man has become less personal and more collective, largely losing his character and individuality in amorphous groups—the so-called "masses." In so far as man has become a mass-man he has lost his sense of the spiritual values—love, truth, wisdom, goodness, beauty. Men desire to hear no more of the demands of a higher humanity, of the duty of spiritual effort and upright living. They want to vegetate in a life of amusement in which a false security is provided by an absence of fear and hope. There is no meaning or purpose in life. Nothing matters.

But more serious than this cultural degradation are the dangers arising from the spiritual vacuum in which the masses live, together with their emotional excitability and susceptibility to suggestion. In their unconscious craving for a purpose in life the masses have formed the

1 This section is composed of excerpts from an article, "Man and Freedom," written in 1942, during the dark days of the war but published much later with slight emendations (ECCLES, 1947). The chosen excerpts seem still to be relevant, even more relevant!, and to fit in at the end of a chapter devoted to Man, Freedom and Creativity. They also have relevance to the themes of Chapter XI, which reveal the tragic and embattled state of our great cultural institutions and the extreme dangers that confront civilized democratic societies as violence, anarchy and moral dissolution well up from the youth.

culture medium for the poisonous germs of the modern pseudo-religions—Nazism, Fascism, and Communism. The deification of race, state or class, absurd in itself, has to them the overwhelming attraction of a this-wordly ideal to believe in and live for. The final result of man's struggle for freedom and erection of his own system is the liberation that turns into total slavery. It is the earthly immortality of man in the form of a collective man, a mass man, a termite man.

It is worthy of our most determined efforts to understand what freedom is. To unthinking people freedom presents no problem. They will tell you that it means absence of restraint, that they are free to do as they please. Such a licence for one individual, however, restricts the freedom of others. Unbridled licence leads to anarchy or social chaos, in which all freedom is lost. The mere granting to all of complete freedom of choice in every act—freedom to be cruel or violent or dishonest or idle—obviously is not the freedom that we have been fighting for.

The freedom that matters is the freedom to know, freedom of thought, of opinion, of discussion. Such a freedom does not limit the freedom of others, and there can be no doubt that it is fundamental, but as well as "freedom to know" we also want freedom in the sphere of action. Freedom to make by our own efforts the utmost out of our lives, to develop ourselves as persons, to give our talents full scope, to live according to our ideals, to control our destiny. This is the freedom that MARITAIN aptly calls "freedom in fulfilment." This freedom too, does not limit the freedom of others. On the contrary, the realisation and progress of the spiritual freedom of individual persons will make of justice and friendship the true foundations of social life. Moreover, the freedom of fulfilment obviously includes the "freedom to know." It embodies all that we mean when we say that we are fighting for freedom.

How can we set about rebuilding our society so that we preserve not only the freedoms we already have, but add to them so as to give the fullest possible life to all men, so creating an order in which all the varying richness of the human personality will be manifested? That is the central political problem of this age. It is not sufficient, however, to provide such equal opportunities. Each person should, by his own will, strive to make the most of these opportunities. Freedom involves not only rights, but also duties. We have no unqualified right to freedom. We are only entitled to freedom in so far as we fulfil the duties of respecting and living up to the freedom we already have. For example, freedom of fulfilment necessitates a progressive conquest of the fulness of personal life and spiritual liberty. Thus, it is evident that we can never attain a static state of freedom. Freedom is dynamic in the sense that we have to be continually striving for it in order even to maintain what we have. Abuse of our freedom endangers it. For example, we endanger our freedom of speech when we use it to make misleading, irresponsible and provocative statements. The right to freedom of speech presupposes the duty of honesty and sincerity.

What sort of society will give each person the maximum opportunity for developing the "somewhat of possibility" that is in him? It is obvious that society exists for man and not man for society; that is, that society must be ordered so as best to fulfil man's needs. The form of society which preserves the right relation of the individual to society and which gives man such opportunities MARITAIN calls an organic democracy, or democracy of the person. At its root we find the idea that man is not 'born free' (independent), but must conquer freedom, and that in the State—a hierarchic totality of persons—men must be governed as persons, not as things; and toward a common good truly human, which flows back to the persons, and whose chief value is the latter's freedom of expansion. Organic democracy is based on justice and on the fullest cooperation of the persons composing it, that is on brotherly love or civic friendship. Civic friendship is not an original state, granted ready-made; it is something to be conquered ceaselessly and at the price of great difficulties. It is a work of virtue and of sacrifice, and it is in this sense that we behold therein the heroic ideal of such a democracy.

Our world is at the parting of the ways. There are only two alternative orders towards which we can move.

One way leads to the centralised planning of the absolutist slave State. In the past a private and personal sphere of life was often able to survive almost untouched by an absolutist tyranny, for despotism was largely devoted to public affairs. With modern efficiency of communication and organisation that is no longer possible. Absolutist governments can only stabilise themselves if they eliminate all resistance before it can organise. Secret police, concentration camps, treason trials and mass propanganda are the inevitable concomitants. Modern absolutism must be total, enslaving man even in his personal and private life. It must be a tyranny characterised by compulsion, terror and collectivisation of every aspect of life. Such slave States may be very stable. HITLER may not have been exaggerating in claiming a duration of a thousand years if he had won the war. The Egyptian tyranny endured for several thousand years. How long will the Russian tyranny endure?

The other way leads by continuous development to the order of freedom and moral responsibility of each human person. It will be an order respecting and nurturing the private lives of all persons, and dependent dynamically on the responsible and free acts of each one of them. At every level of society there will be full play for responsible action. In other words, one of our first considerations will be the widest possible extension of personal responsibility; it is the responsible exercise of deliberate choice which most fully expresses personality and best deserves the great name of freedom.

How can this fostering of universal responsibility be linked with the demand for efficiency that is so essential for survival in this modern world? Here is a field for constructive thought and experiment both in education and in political action. We may well ask, is it true that central bureaucratic control always gives the maximum efficiency? Surely in most affairs of social life the co-operative action of responsible persons should be more efficient than the passive obedience and subservience of slaves. Co-operative and responsible action provides that tension in living, that continual process of give-and-take, in which we are stimulated to give of our best, rising to heights of achievement otherwise impossible. Does not history show that our cultural heritage has been built up by the responsible acts of free men? And yet is not the shirking of responsibility one of the most widespread and dangerous symptoms of our age?

The fostering of responsibility is thus one of the most vital tasks of reconstruction. Society must be conceived as a hierarchy of functional groups with the surrounding responsibilities. We have firstly the family with the individual responsibilities of its members; secondly, small social units such as neighbours, workmen and employees in a factory, office or warehouse, as well as groups combining for religious, cultural, scientific or charitable activities or for recreation; thirdly, larger units subserving civic responsibilities and social services; then regional councils and industrial councils, and so on to the supreme governing authority. The distribution of responsibilities is determined by the principle that every function which can be assumed by the inferior must be exercised by the latter, under pain of damage to the entire whole. In all questions outside foreign policy, defence and general legislation, political decentralisation should be pushed as far as possible.

During the last few hundred years scientific investigation has gained for man an understanding of nature which has given him an astonishing mastery over it. During the same period there has been no obvious advance in man's knowledge of himself as a human being. There has been much theorising, but no steady progress in factual knowledge. Sociology, economics and politics are not yet sciences resembling the natural sciences. It is extremely difficult to apply that same scientific method to social and political problems, for we ourselves are the raw material we would manage. To that material we have not applied the scientific method, with the results that we see in the present state of the world. Man is unable to control himself in the way that he controls natural processes. The many doctrinaire plans for building a Utopia are reminiscent of the philosopher's stone of the alchemists.

We have as our goal a society which gives each man the optimum conditions for leading a full and responsible life with the full development of his talents, and not, for example, a society for the maximum production of wealth, or for the maximum development of trade, or for the maximum concentration of power in a governing oligarchy with the maximum subservience and passive obedience of the people. By the use of scientific methods we should gain useful knowledge of the way in which we may develop our society so that it approaches ever closer to our ideal society. We have to learn to apply to ourselves the slow, laborious technique of searching for truth by which science has built up knowledge of natural phenomena. Of course, up to a point, the scientific method in its empirical aspect has always been applied haphazardly to the ordering of society. What we still largely lack is the systematically planned experimental investigation with proper controls and unbiassed interpretation. Doctrinaires have a theory of politics or economics and try to fit man to it. But politics and economics have no absolute values in themselves. They must be judged by the criterion of how far they contribute to the well-being of man living in society. Man must be the measure of the system.

Before setting out on such experiments it is fundamental to realize that our civilisation is based on the accumulated wisdom of the past, on the empirical discoveries of countless individuals. We must build for the future on this great culture and tradition of the past, modifying it as little as possible. The essence of progress is not liberation from the experience and the standards of the past, but their revision to meet the needs of the present. Culture and civilisation are not synonymous only with progress; they comprise also the forces of preservation; they imply, also, the continuity of the age-long ascent of man.

Chapter IX

The Necessity of Freedom for the Free Flowering of Science[1]

I have chosen this title because it is consonant with the magnificent citation for the Dunning Trust Lectures. I do not know of a better citation for any lectures. Let me quote: "To promote understanding and appreciation of the supreme importance of the dignity, freedom and responsibility of the individual person in human society." I have really struggled with this lecture because I was inspired by these words, and that I think is the best tribute I can give to the words. Furthermore, it corresponds so well to my vision of what can be claimed to be the greatest spiritual adventure of our present civilization, an adventure that is subsumed in one word, Science. In Science the freedom and responsibility of the individual scientist are of supreme importance.

Science and Technology

I would define Science as the systematic attempt to understand and comprehend the natural world. It means to enter deeply into the natural world, not just superficially, but to achieve a rational expression in language and mathematical symbolism of the order and beauty of operation that lies behind all natural phenomena. This is what the aim of Science is. Its scope is not just the external world. The scope of Science includes ourselves. By that I mean that all aspects of our own experiences of ourselves—our perceptions, imaginings, emotions and actions—everything properly comes into the purview of Science.

Now this enterprise of Science, is dependent on the imaginative and disciplined activities of unique individuals who ideally form a free and supranational society. This essential feature of Science certainly is not fully understood and appreciated by the politicians and the bureaucrats that operate even the most liberal political systems, and in most countries it is not understood at all. Correspondingly there are

1 Dunning Trust Lecture given at Queens University, Kingston, Ontario, Canada, on October 15, 1968 (ECCLES, 1969c).

not very many countries in the world where Science flourishes in a free-flowering manner, as I like to describe it metaphorically. There are far too many countries where there is no growth of Science at all or where it is extremely maimed in its growth. I had not fully recognized the impoverished state of so many countries until a world survey of brain science was done for the International Brain Research Organization of UNESCO.

In some countries of up to a hundred million people, there are no brain scientists at all. There are many countries, offering in general, unfavourable conditions, but with isolated growth and flowering, that are dependent upon the courage and genius of one leader and his devoted disciples. As an illustration from the past, I give RAMÓN y CAJAL, who late last century in Spain (which was completely unfavourable) and with practically no support at all, became the world's greatest neuroanatomist, building up a world famous school and making a magnificent scientific contribution that to this day we are still so much dependent upon. There has so far only been one RAMÓN y CAJAL in the sciences of the nervous system. He exemplifies a very rare phenomenon; the high level of scientific performance that can be achieved by a genius under unfavourable conditions.

Before I enquire about this extreme diversity between different countries in respect of Science, I will draw a sharp distinction between Science and Technology. I am speaking tonight of Science, not of Technology. I am not going to say which has the higher status, but it is necessary to be clear about these two categories, though the same person can be both a Scientist and a Technologist. He can wear, as it were, different hats for the occasions, but what he does on these two occasions is quite different. There are technologists in many countries where there are no scientists. In fact all countries have technologists even when they are in the Stone Age. Then they use Technology for fashioning stone tools! In Professor WASHBURN's School of Anthropology in Berkeley, California, the graduate students in Anthropology, as a practical exercise in the archeological laboratory of Dr. DESMOND CLARK, spend a whole semester in trying to make a stone tool, using the same tools as Stone Age people had. Professor WASHBURN (1969) states that no one so far has succeeded in making a stone tool that is in the class of those made by large brained men some hundreds of thousands of years ago! However after a long intensive course it is possible to duplicate the ancient performance.

The aim of the technologist is to apply scientific knowledge and empirical knowledge in some useful purpose. He may even be designing and constructing scientific equipment. I am a technologist when so engaged. Technologists are responsible for all the marvellous inventions that have transformed the conditions of our life. You only have to think of a spectrum ranging over the revolutions in the means of communication, in the materials for all purposes, in electronics, and all the new discoveries and inventions relating to medical practice, to agriculture and food, to chemical industry, to computers, and so on. This human activity is quite different from Science, although it utilizes or even exploits the discoveries of Science. Of course, the technologist has to have a wide knowledge, imagination and high intelligence, just as does a scientist. I am not underrating the level of his performance, I am just saying that it is different from that of the scientist. The difference lies in the different objectives.

The scientist tries to understand or comprehend the natural world as he experiences it, using for this purpose hypotheses that are tested under specially designed experimental conditions in which often most elaborate scientific instruments are employed. Yet, in the end, no matter how complex the apparatus is, the information that it delivers has to be looked at and observed by a scientist, who has to take notice of it, and critically examine it in relation to his hypotheses, and maybe reformulate his hypotheses and so on. The essence of what a scientist does is to imagine and explain what lies behind phenomena, in other words to try to comprehend the natural world. On the other hand, the technologist is utilizing the knowledge about the natural world for practical purposes. That statement applies not only for all the practical things that I have spoken about, but it also applies to space travel, which is really Technology. It is marvellous engineering, but it is not Science. It is space exploration. There may be a little of Science in it, but it is not a scientific achievement; it is a tremendous technological achievement utilizing the discoveries of Science for its purposes. Another side of Technology is of course the threatened horrors of atomic warfare—target-finding missiles with thermo-nuclear war-heads.

It is a tragic mistake that the general public and the political leaders confuse Technology with Science, there being virtually no appreciation or understanding of Science itself. I can give you two examples. The belief that space travel is an advanced form of science can lead to most embarrassing situations. I actually made reference to this misunderstanding in a lecture I gave in Melbourne while the first Sputnik was in orbit unbeknown to us. Next day I had many interviews with reporters as I was then President of the Australian Academy of Science. A few days later the Soviet Ambassador to Australia asked me why I did not indignantly deny the absurd statements attributed to me. He stated that "Do you not recognize that the orbiting of the Sputnik is the greatest scientific event since the discovery of America by Columbus?" To which I replied, not very diplomatically: "Your Excellency, I am greatly surprised that you think the discovery of America had any scientific significance whatsoever!" I received no more invitations to the Soviet Embassy. Only a few days ago (October 1968) it was reported in the Journal of Scientific Research that an eminent political leader in emphasizing his support for science had stated "I understand that recently a gap has developed between Soviet and American Science. My efforts will be to narrow that gap!" This statement has a most unfortunate implication when one appreciates the very large gap in the inverse sense between the levels of the true science of America and the Soviet!

This is the fate of the most scientifically advanced country today, that its leaders have very largely been misled on what Science is, and this defect will be recognized in many of the points that I raise for discussion. Soviet Russia has been credited with high scientific performance because of this misunderstanding. They have been able to peak in a few Technologies, space travel and atomic missiles for example, but their performance in almost all aspects of Science is quite miserable. The exceptions are some branches of Physics and Mathematics.

The mistaken identification of space travel as the most significant Science of our day has the further unfortunate consequence that science will be discredited when space travel fails to deliver a continuing stream of startling discoveries. The general public and even the most enlightened political leaders fail to appreciate the extraordinary conditions that must prevail if there is to be a free-flowering of Science. It will be my task in this lecture to show that of its very nature the scientific enterprise depends on a very special freedom of the scientific investigator.

Science only came to our western civilization in the 16th and 17th centuries. Before that there were only brief glimpses of it, and since then it has gone on by fits and starts, flourishing and flowering or just wilting. The 17th century was a century of development; the 18th century, much less so; and the last half of the nineteenth century was a period of great achievement that has gone on at accelerating pace in this century. That brings me to enquire what special conditions are required for there to be a rich scientific achievement in any age.

The Making of a Scientist

Let me begin by enquiring how a scientist comes to be. I will consider this sequential development in five stages, and on this basis attempt to explain how it comes about that some ages and some countries are distinguished by having many scientists and a flourishing Science, whereas most fail. In this enterprise I am specially indebted to Professor POLANYI's most illuminating books (1958, 1964, 1966).

(1) Our civilization has a naturalistic as opposed to a magical outlook on nature. We do not believe in such influences as the occult, in astrology for example, though of course, there is always a lunatic fringe! Instead the child grows up in an atmosphere in which he learns to believe that there is a rational explanation of everything he experiences. A child will ask questions of his parents or teachers and often they are very good questions. Thus he learns that nature is rational and orderly and capable of being understood. Almost every child in our community grows up in this atmosphere, thinking that, even if he cannot understand it, there is at least a good explanation. For example he is offered explanations of the movements of the sun and moon, of the tides, of rainbows, of streams flowing and of the origins of rocks and fossils. The world is orderly and capable of being understood at least in principle. He has an example of what can result from understanding, because we have now progressed so far in Science that there have been immense developments in our control of nature, as witnessed by all the wonders of Technology, such as machines for every purpose, and the child-like wonders of radio and T.V. The message is certainly compelling to impressionistic youth.

(2) After that initial education, some young people, particularly the inquiring ones, say: "I would like to get an idea of this for myself. It is not enough to be told these things. Can I go on?" So he may elect to do Science here at Queen's University, or some other University, where he undertakes the ordinary undergraduate course. Here he gets basic instruction in scientific discoveries and in the laws that scientists have established. However, he does not really understand Science at this undergraduate level. I think he is only just learning about Science; he is not learning what it is to be a scientist; nor does he learn what motivates a scientist, or how a scientist behaves in making discoveries, or in fact what is behind the whole process.

(3) Our embryo scientist may then go on at the postgraduate level if he has been successful and done his homework. In parenthesis I can assure you that I have been through all this myself. I was extremely successful at examinations so that I got a Rhodes scholarship to Oxford, but still I did not know what Science was about. I went there and learned. I learned that you do not get clear answers, and there are many arguments, schools of thought, changes of opinion and so on. I began to learn much more when I started out in collaboration with RAGNAR GRANIT to do research with Sir CHARLES SHERRINGTON and with the critical and wise advice of DENNY BROWN, who was my senior in research by two years. The kind of thing that is learned in the Ph. D. course is that Science has an eternally provisional nature. It is uncertain and so has unlimited possibilities for growth and change. One must cease to believe in ultimate statements which claim to give a finally established truth that should be

accepted unquestioningly. Let me repeat that basic scientific ideas, theories, and explanations are always provisional and always subject to change. For example, think what happened to the magnificent Newtonian theory. In this stage, the Ph. D. student gets some routine experience in research. On the whole most postgraduate students are given fairly tidy little jobs that are planned so that ordinary laboratory equipment can be used. They are supported on all sides by previous scientific investigations, so it is improbable that anything startlingly new will turn up in this research; nevertheless it is worth doing because it fills in gaps and makes the conceptual structure more complete. If our student is lucky, he may come across something that was unexpected. CHARLES BEST at the age of 21 or 22, with BANTING, discovered the Insulin story, nearby in Toronto. This can happen when he is very young indeed, but it usually does not. However, he will have learned quite a lot about how to do experiments, how to observe and how to handle abnormal observations when they come. There may have been something wrong with the animal, with the equipment, with the substances such as chemical impurities or a thousand and one other things, or there may be the first inkling of a new discovery!

(4) If our young scientist is good, he should go further than Ph.D. level, and in postdoctoral research actually experience creative Science both conceptually and experimentally. At this stage it is important to pick the right man to work under. There is no textbook which tells you how to become a scientist. There are lots of books about the methodology of Science and also the philosophy of Science, but they are mostly wrong and misleading. A scientist has to learn in the old-fashioned way, and be an apprentice. It is just like in the Renaissance times if you wanted to be a painter. For example, you would go to the studio of Verocchio, where you learned about pigments, anatomy, perspective, technics of fresco and all those sort of things, and you were allowed to paint here and there on pictures. Eventually, you may have become a Leonardo da Vinci. Leonardo was apprenticed to Verocchio when he was 14 and worked in the studio for 6 years. This is essentially the way that Scientists are made today.

In my experience the relationship in a laboratory cannot appropriately be called a Master-Pupil relationship. I just work with people. For example, I arrange to do some experiments with them usually on some problem that has arisen in a preceding investigation, but that has not yet been solved. During the experiments we watch the results together and we talk about the literature. We try to interpret what we observe in the experiment, going backwards and forwards about the meaning behind it—"Does that strange phenomena mean anything, or does it have no significance, merely arising on account of defects in the experiment?" In that way one's scientific associates can be given an insight into the personal intuitions, hunches, serendipity, which somehow one has learned and which one delights handing on. Then always in trying to solve one problem we discover new phenomena that raise more problems, some of which we attack in turn and so on.

Science is an art. It is an art because it has to be learned in this strange manner. It is a way of looking at things, of seeing behind them, a way of imagining. Creative imagination is required, and in this kind of association in a laboratory there is one of the most satisfying experiences in human living, a real delight in the exchange of ideas. However it can be very tough at times; you fight about meanings and ideas, but it is not a one-way action. I can learn, and in recent years I have learned a great deal from what you would call my junior colleagues. I have been wrong in my explanations and of course eventually recognized that I was wrong. That is the way it goes. Sometimes, they are wrong too! I would develop an awful inferiority complex, if I were always wrong! It is this disputation backwards and forwards which is the essence of Science, and it is in this way that you can create a scientist. There is the exception, like RAMÓN y CAJAL, who created himself, but this is asking too much, even for the extraordinary person.

You do experiments and you can have a very good idea of what to expect from your past experience, and your expectations may be fulfilled. Usually, however, there is something else coming in, some worrying phenomenon and for a while, you take no notice of it. You even try to look aside when it appears! But it keeps on reappearing, so you recognize that "here is something that Nature is trying to tell me." This is essentially the way in which I have made discoveries. They arose from the happenings that I was not expecting, where eventually I was tuned in enough to listen to what Nature was trying to tell me. The good students and the good research workers are those who can recognize and appreciate the significance of the unexpected. Those that you discourage and tell to devote themselves to practical affairs are those who, when asked how they got on with an experiment, say that they were terribly disappointed because what you had predicted had not happened. They did not recognize that, when confronted by the unexpected, you really should be alerted; and this is what a good scientist has to learn. This is what he should learn in a good environment.

But after much experience I have to admit that there is also a perversity in some students that lowers one's assessment of their quality as potential scientists. As one learns from one's own hard experience, wisdom and judgment are required in the interpersonal relations that are so important in the creation of a good scientist. It is good to be motivated by ambition, but those dominated by ambition become impossible colleagues, and even their scientific integrity may suffer. The overly ambitious irritate by their arrogant posturing and display.

(5) And now comes the final stage of being able to lead an adventure and to have pupils in this enterprise. After two or three years post-doctoral experience with working under good conditions some young scientists can venture on their own and succeed. It is most encouraging when you find that your research associates develop in this way from strength to strength, making remarkable discoveries. I had with me for some years in Canberra Dr. MASAO ITO who went back to Tokyo to develop his own research group. They performed experiments on neural pathways from the cerebellum that led to ideas which I just could not possibly believe when I first heard them. They were so original that I just rejected them, but eventually I had to realize that this was a new and quite fundamental discovery, namely that the whole output of the cerebellar cortex was inhibitory. I regard it as a great success, when your pupils graduate to such an amazing scientific performance.

The Discipline of Science

Now I come to what Science is. Hitherto I have been trying to give you a feeling of the free society in which a scientist is created. It is neither a dogmatic society nor an authoritarian society. It is a free society of people adventuring into the unknown. Following Sir KARL POPPER (1959; 1962) I can state that what we do in Science is essentially that we have ideas and hunches going quite beyond anything that is yet established as a result of experimental investigation. These ideas or hypotheses offer challenges which become formulated as predictions capable of experimental tests. Scientific hypotheses should be defined so closely that they challenge attack. One of the best compliments you can be paid is that someone spends days and even years trying to disprove some hypothesis that you have proposed. You have challenged them and maybe they will disprove you, but this is Science. It is a scientific success to have disproved a good theory. Science advances in this way, refining and clarifying ideas by rigorous testing. As a result a hypothesis can be said to be corroborated, but it should never be claimed that it has been proven true. Ultimate truth is still beyond that. No scientific hypothesis or theory today can be claimed to be completely true in itself. We must say it is the best we can do. You have an example of this in Newtonian Mechanics, which for more than 200 years was regarded as the ultimate truth, but now of course, you know it is only a good approximation. This is what will happen to all our theories of today. Some of us scientists have become very wise about this, but non-scientists do not understand it at all; they think that Science gives truth, and that, they think, is the great virtue of Science. This is also believed by most scientists today. That is the trouble. They believe in a false philosophy because they have been badly brought up theoretically. As long as they have been well brought up in the "practice" of Science, they can get along quite well, doing things correctly, but misunderstanding the essential nature of scientific investigation. They believe with BACON that Science consists in observing nature, and when you get enough of these observations, you then get from them a pure distilled truth. You collect a lot of grapes in a barrel and then you sit on it and out comes the wine of truth!

There are many other strange beliefs associated with this so-called inductive theory of Science. It is erroneous, because you are just misleading yourself if you think that observations in themselves have scientific significance. In fact you only observe special happenings because you have some ideas about what to look at. You just cannot look at anything. I can make millions of observations in this room, but they have no scientific value whatsoever. A scientist does not just make observations, he makes observations within the special framework of a theory which gives him highly specific information about what to look for. It is also a mistake to believe that nature cannot lie, and that, if you listen with an unbiased mind to what nature has to tell you in response to your questioning experiments, and write it down, then you have a truth. The whole point about it is that nature would only give you the most confused stories – nothing at all scientific, just myriads of happenings. Instead a scientist has to develop from his hypotheses specific experimental situations by which they can be subjected to crucial tests. By experiments that are well designed and carried out he is asking carefully framed questions of nature, and discovering how his hypotheses stand up to this experimental scrutiny. That is the way scientific investigation is carried out. In contrast many of the so-called scientific publications are just compilations of observations often made with the most sophisticated and expensive equipment, but seem to lack any scientific meaning.

It is important to appreciate that inevitably we have tensions between innovations and the orthodoxy of the Establishment. A scientist must be an innovator. Previously I mentioned MASAO ITO, who discovered something quite unexpected, fundamental and new, and my reaction had immediately been to reject it, and say that something was the matter with his measurement of time, or that his electrodes were not properly recording, or something like that, because evidently I belonged to the Establishment; yet eventually I had to recognize his discovery. All the time in Science there is the Establishment, composed of those holding orthodox views, and on the other hand, the innovators, the people making discoveries. A great many of these claims to discovery by innovators are, of course, phoney.

How then do we manage to progress in Science? Well, a scientist does some experiments, and gets a lot of results. He writes an account of his problem, his results and a discussion of the results. He sends his paper to a journal, or he could publish a book. The ordinary way is that you go through the mill of presenting your data to a scientific journal. The journal has editors. I am an editor of two journals and also an advisory editor of many, so I am often called to referee papers, as it is called. Our task is to criticize and evaluate the work, and also to make suggestions about changes that should be made if it is to be accepted for publication. If you are unable to get your paper past the referees of one journal —and even now I have difficulty in getting some of mine past—you can try another journal. There is not just one journal, with one set of editors, there are many journals and you are free to pass your work around, just as long as you do not publish the same paper in two different journals. You can try in succession until one will take it, and as far as I can see almost everything gets printed eventually! Often papers have to be modified and usually referees give good advice for revision.

Once the paper is printed, the next stage that the scientist has to go through in his testing is that it is read, and sometimes by experts in the field. If you are an expert in the field, at first you look it through very cursorily and see if it is worth reading, and secondly you decide how much time you should spend with it. Sometimes, you read and re-read it, making copious comments often in the margins, and sometimes it just disappears from your mind and even from the literature. Some papers have an extraordinary short "half time." I am told that in America there is an organization that for Presidents of Universities and such like will carry out investigations on any named person to find out how fast his papers disappear from the quotational literature! Sometimes it is too revealing. It is all done on computers within a day or so, and at a price you can get the answers for all candidates applying for a job! Frequently before or during the course of publication the scientific results and ideas are presented verbally at some scientific meeting, where the author has the opportunity of defending himself against criticism or of further explaining his results.

The next stage may be even more remarkable. You may find people not only quoting your paper, but also accepting it so that they base their own experiments upon it, or they may even write textbooks or monographs with the paper occupying a key role. This is the ultimate success. However, don't be misled by that. It still could be wrong, and probably is at least in part wrong. It takes longer to get something out of textbooks than it does to put it in. But even after this success stage it still remains open to attack—which may either be directly in new experimental developments or it may be when it is found to be inconsistent with quite other investigations. And so the scientific process goes on—an interplay of conjectures and refutations as described by Sir KARL POPPER (1962).

If your paper is rejected by a scientific journal, you don't have to be disheartened altogether. The distinguished scientist-philosopher, MICHAEL POLANYI, recently has written (1967c) advocating strong editorial policy, rejecting all papers which look unassimilable to the present scientific beliefs. Otherwise, the ones that appear significant, the ones that look to be developing in the right direction, will be choked by the rank growth of an immense number of misleading papers that would fill the journals. He cites a case of his own. Many years ago when he was a young scientist his paper on the "Absorption of gases on solid surfaces," was rejected by the editors of a journal. Although this paper was not published, there was a record of it, and thirty years later, it was found to be correct. It happened to have been far ahead if its time in describing a quite new and unexpected development, and as a consequence was rejected by the Establishment. A few weeks ago I heard MICHAEL POLANYI cite this case and defend the editors who had rejected his paper forty years earlier when it did look unacceptable to them. This is an example of the enlightened and tolerant attitude that scientists should have towards the editors of scientific journals.

I think it is very important that we should cherish and appreciate the operational aspects of scientific societies and journals. The tradition is that the scientific societies do not intrude on the scientific freedom of their members. You have to realize that as august a body as the Royal Society of London does not adjudicate as to whether any particular scientific statement or hypothesis is true or false. Its members can do as they like, but it is the tradition that as a society it never adjudicates. The same would be true for the National Academy of Sciences, and so on.

Scientific orthodoxy thus is not maintained by some formal establishment. It is dependent in a strange manner on the informal and often unorganized discussions between small groups of scientists. That's how Science goes on. We are all the time chattering, gossiping about this or that scientific story or scientist—the incessant gossip of the Laboratory. In fact laboratories should be maintained for this purpose even when there is no experimental program! Two or three days ago, I took a paper into my laboratory, and I said, "Here is a paper on the subject we are working with. This work could have been done last century, and even then it wouldn't have been too good." That is how you establish standards with your research students by showing not only good things, which I do regularly, but also the bad, and letting them see how you evaluate scientific contributions, both experimental and theoretical. They have to learn from this general experience about an on-going scientific discussion that can be quite searching and merciless. Scientific meetings are often most valuable in the opportunity they give for free informal discussion between the program items or even during the items, because, except at very small and specialized symposia, absenteeism is much practised—often to advantage! Unfortunately the programs are usually much too full, but even then there is hope if some Russians have been invited. Many conferences that I have attended have been opened up by the holes left in the program thanks to the tyranny of the Russian bureaucrats in vetoing attendance by Russian scientists!

Scientists can be very hard in their criticism of each other, but they also, of course, match that with generosity and admiration. The level of performance is always being assessed and judgements may often appear to be uncharitable. However, at least it is rare that accusations are made that a scientist lacks integrity. Criticism is not made on those terms. Rather it is that he is misguided and has mistaken ideas or even not worthwhile ideas at all!

Freedom and Science

Now I come to freedom. I have talked about Science and the atmosphere in which it is conducted, now I ask: how does freedom fit in with this? Why is freedom, as we define it, so necessary? Its necessity comes from this tension between the two different themes of Science. On the one hand there is the freedom to adventure idealogically and imaginatively and so to have the opportunity to experience new visions in the understanding of nature. All scientists except those conditioned by their political masters will believe that freedom is necessary for this. There must not be some imposed dogmatism limiting what can be done or expressed in Science, as for example, happened to Galileo. On the other hand, I have referred to the critical and even repressive activity, which is the general authority that is exercised by the scientific community with its scientific societies and informal groups. They function as arbiters, but are not properly or even deliberately constituted for this purpose. There is the on-going discussion and criticism all the time by scientists whenever they meet. It is the tension between these two opposite poles of scientific activity, on the one hand freedom to adventure, and on the other critical restraint and even repression that gives Science its amazing strength and success. Sir KARL POPPER (1962) uses the words—conjectures and refutations—for these two opposite poles (cf. Chapter VII). Very rarely does anything phoney survive for long. Scientists may develop and express mistaken ideas, but experimental refutation soon prevails.

In recent times, Science has not gone widely down any side tracks, not the main Science. If you want an alternative you can think of the genetics of LYSENKO in Russia. I am not advocating unrestricted investigations in the name of Science. I like to think of Science as a great garden, replete with wonderful plants, but also threatened always by the concurrent growth of weeds. By the criticisms at meetings and refereeing for journals most of the weeds can be eliminated or at least controlled from proliferation, but at the same time wisdom and insight are required so that strange new fruitful plants are cherished and not pulled out when they are immature and hard to recognize. When a scientist can grow a beautiful new plant that eventually develops in a magnificent free-flowering manner, then that brings, of course, the highest rewards in status from the scientific community.

I think the safeguards against arbitrary action are given because in a free society the community of scientists does not act as a single dogmatic authority. Instead there can be many authorities, often advocating different and even opposing schools of thought. There will be many mistakes, but usually they are not fatal to any individual. And there is no rigid dogmatism in most fields. Thus freedom of scientific exploration is safeguarded fairly adequately. In the free countries of the West there is also freedom to travel from one country to another for scientific meetings and for scientific training. In fact it is greatly encouraged by generous financial provisions.

If the young scientist we have been imagining has ideas that do not fit in with the professor he is working under, he is free to move to another place where the scientific ideas are more consonant with his own. He can move because he has freedom to travel, which is very important, and also because there are always places for dedicated scientists. This is why a free society is so important. Even within most laboratories there is much room for differences of opinion and even a generous tolerance of dissent.

This freedom to travel internationally does not obtain in the Soviet Union, where only the safest people are allowed out, particularly for an early period of training, and where the young usually work where they are told to work and under the master to whom they are assigned. Their fate is to continue with experiments further confirming the orthodox views of the master. It is pathetic to witness the scientific presentations of some of these unfortunate people, young and clever people, who are working in this system. Russia is the most reactionary country in the world, and the Kremlin exercises an authority that is unbelievable in the free world. Much more freedom is still preserved in the so-called satellite countries which are really colonies in the Soviet Empire. I had a young scientist from a satellite country working with me some years ago. He appreciated our free society very much and after some months of observing and thinking he said, "I think I now know the difference between the East in Physiology and the West in Physiology. In the East, I was taught that you had to get some ideas which were orthodox and official and you had to do experiments so as to support and confirm these ideas. Here I find out what you are doing is: firstly you have some general ideas that lead to experiments, then you seem to listen to what nature is trying to tell you, so that you can develop better ideas." That, of course, is what Science essentially is.

Now, I return to the question that I raised at the beginning. How does it come about that some countries have been so successful in Science? If you will allow me, we will take as a test for Science, the Nobel Prize Awards. The Swedish Academy of Sciences and the Medical Faculty of the Karolinska that make these awards are very well acquainted with Science as distinct from Technology. They are giving the awards for Science in almost all cases. In December, 1967, when Britain was at a low economic and political ebb and everyone was deriding the deterioration of Britain, Sir PATRICK BLACKETT, President of the Royal Society, in his annual address (1968) made some startling statements based on arithmetic. He added up the number of Nobel prizes that had been given in the post-war period, and found out that the United States had the most. It had 54, but it was surprising to find that Britain, with only one fourth of the population, had 24. At that same time, the European Economic Community, which was flourishing economically, Germany, France, Italy and the Benelux countries, could only muster 15 Nobel prizes. Soviet Russia had only 7, none of these being for Soviet Biology and Medicine.

Soviet biology is really very poor. To give an example from basic neurophysiology, I wrote a book some five years ago. In it there were two citations from Russia in over 1,000 references. In my two more recent books there was none! However, there were a great many references to work done in the so-called satellite countries, where there is still freedom, despite the Russian efforts to quench it. Recently I have asked several experts in the neural Sciences how many Russian references they would put in a monograph, and they usually said, none. I have also counted the number of times Russian papers have been quoted in recent literature. In 1968 in the international journal, Experimental Brain Research, published in Germany, there were only 7 references to Russian papers out of a total of over 1800 references. In the international Journal of Neurophysiology, published in U.S.A., there were in 1968, 6 references to Russian papers out of a total of 2218. This is the kind of Science that our political leaders talk about. We do not understand the level of performance. I am told that some aspects of Russian physics and mathematics are much better. In those subjects the scientists have been able to free themselves to some extent from the ideological strait-jacket.

Another country which suffered greatly during the war is Japan. The recovery has been amazing. In my book of five years ago (ECCLES,

1964) there were 47 citations of Japanese papers, and many of the illustrations were from Japanese science. In a more recent book published two years ago a Japanese and a Hungarian were my co-authors.

Now why is Britain so good in Science? I think it has the right idea about Science. You will recognize that I greatly admire Britain, and I am happy to talk like this in Ontario! If you have attended the British Physiological Society, you will know what a scientific community is like. You have to belong to that Society and see the meetings, hear the discussions, see the friendliness, the criticisms and experience the whole atmosphere; and you realize that here is superb training ground for scientists. The same is true with their Chemical Society. As judged by Nobel awards, British performance in chemistry ranks very high in the world. For example, to date it has second place in Nobel awards for Chemistry —17 as against Germany's 22 and 16 for U.S.A. These records give good examples of the high level of British Science. The British are not nearly so successful in the application of their Technology—that is the trouble. However Sir PATRICK BLACKETT pointed out that, despite its relatively poor economic position, Britain was spending much more on research and development than countries of the EEC, so he predicted with confidence a recovery of Britain.

Where there is a slavish copying of the masters that control the Establishment, Science wilts and fails, as has happened so often in past ages, and as happens today with biology in the Soviet Union. I am not saying that the Russians are less intelligent or less able to be scientists. In the Czarist days they did very well, and in many fields led the world in biology, PAVLOV being a notable example. All his work that matters was done in Czarist days. They have failed in recent times because they have had the Establishment of LYSENKO and they still have the Establishment of PAVLOV's ghost operating over great parts of Physiology and almost defining what has to be done. They have performed extremely well in some branches of Technology because the dictators can dictate policy and can channel funds and practical prestige into these problems. So far as I know there is almost no Science coming out of China, except in special channelled fields, and no imaginative insights. So, I would say that you have only to think of GEORGE ORWELL's (1949) prophetic book, "Nineteen Eighty Four," and you can see how by double-thinking truth can suffer in states that have lost integrity and which engage in the deliberate propagation of lies by the fabrication of "truth." You will see an example of ruthless repression in the recent trials of those five people in Moscow for a very mild demonstration in Red Square. There are of course clandestine expressions of discontent with tyranny and the advocacy of freedom by writers and scientists in Russia such as SOLZHENITSYN and SAKHAROV, but SAKHAROV's writings cannot be published

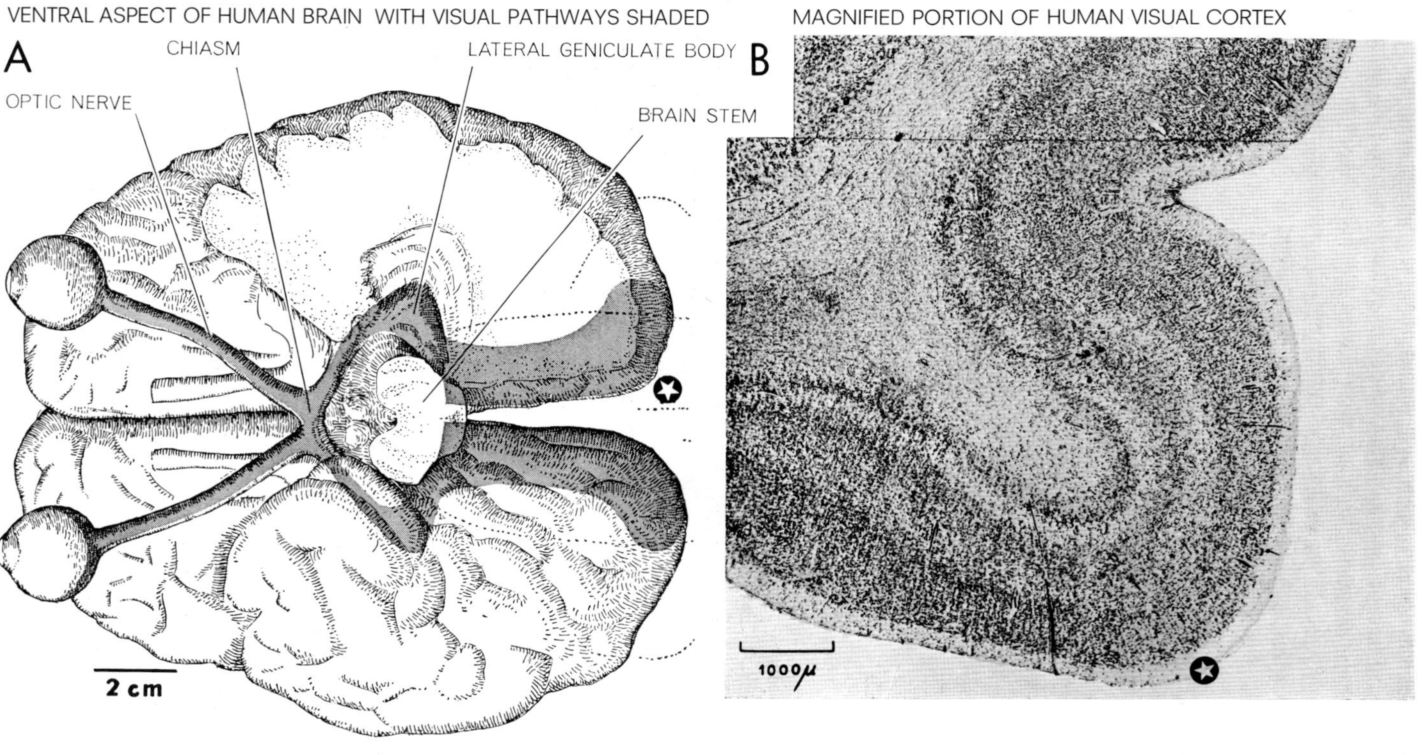

Fig. 31 A and B. Neuronal structures in visual perception. A Ventral aspect of human brain with visual pathways shaded (HUBEL, 1963). B Magnified portion of human visual cortex. White stars indicate similar portions of the cortex (SHOLL, 1956)

conventional two or three dimensional arrangements, but in an N-dimensional arrangement where N (the convergence or divergence number) could have a magnitude of even hundreds. Here is a challenge for multidimensional geometry (cf. Chapter VIII).

It is now well established (cf. Chapter II) that there are two kinds of synapses in the cerebral cortex (cf. Fig. 32C), one kind is excitatory, and, if there is sufficiently intense bombardment by excitatory synapses, the recipient nerve cell can be stimulated to discharge impulses along its own axon (Fig. 5B) and so it becomes an effective unit in the multidimensional network. The other kind of synapse is inhibitory. Inhibitory synapses counteract the excitatory synapses and tend to silence the recipient nerve cells (Fig. 5J, K). Thus each nerve cell in this immensely complex multi-dimensional network is continually subjected to bombardments by the excitatory and inhibitory synapses upon its surface and its responses derive from the net effect so produced on it. There has been much detailed study of the synaptic structures upon the surfaces of the dendrites and the somata of nerve cells and of the manner in which their axons branch to achieve distribution to other nerve cells. However, as yet our conception of the cerebral cortex is in terms of the behaviour of individual cells and of the interactions of very limited ensembles of cells (Figs. 2, 3, 4, 7, 8).

It has been shown (cf. Chapter II) that the essential time of action of the nerve cell in an operative linkage is about 1/1000 of a second for cells with short axons. This represents the total time between the reception of the synaptic excitation that triggers the impulse discharge and the action of this impulse at synapses upon other nerve cells. It will be recognized that, once activity arises in a population of nerve cells, it is potentially capable of almost explosive spread throughout the neuronal network—to millions in a few milliseconds, but inhibitory synaptic activity mercifully can restrain this explosive spread, which otherwise would result in a convulsion. It is still impossible to conceive the manner of operation of the neuronal network operating in a more global manner and involving tens of millions of neurones, which of course would be occurring in the cerebral cortex during all kinds of conscious experiences such as memories, thoughts, dispositional intentions.

More success has attended studies of the manner in which the brain is "informed" of the goings-on in its external world, which may be external or internal to its body (cf. the above introductory citation from Sherrington, 1947). For example electromagnetic radiations with wave lengths of about 400 to 700 mμ are transduced in the retina to give discharges of nerve impulses along the optic nerve fibers. These impulses are brief all-or-nothing electrical events that travel without cross-talk along the million or so nerve fibers in an optic nerve, the pathway being

NEURONES AND SYNAPSES
A NEURONES VISUAL CORTEX

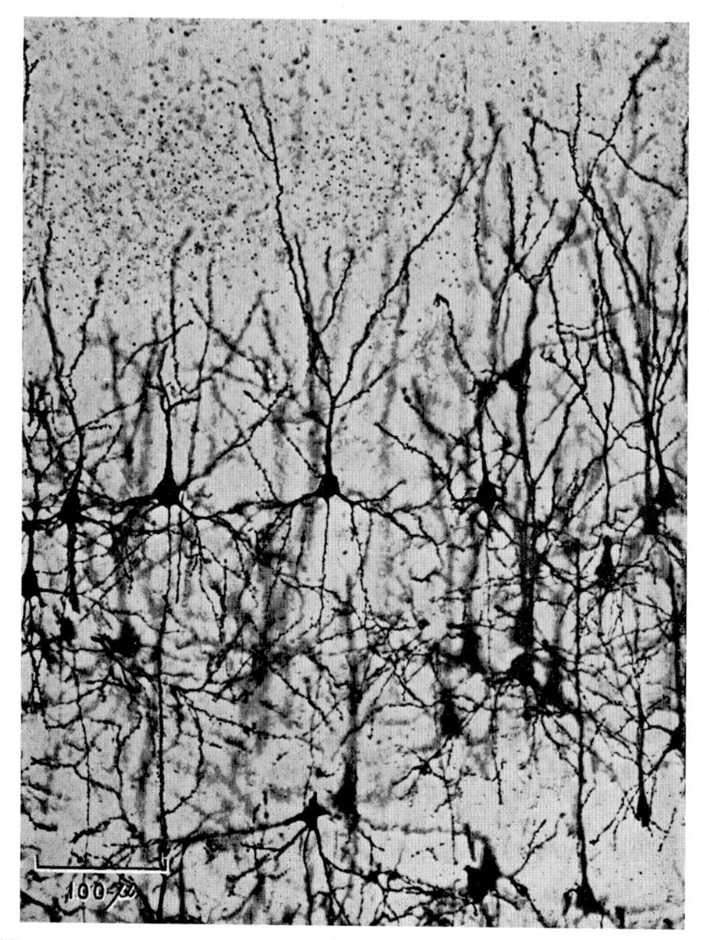

Fig. 32A–C. Neurones and synapses. A shows pyramidal cells of the cat visual cortex (SHOLL, 1956)

indicated by shading in Fig. 31A. The information from the retina is transmitted to the visual cortex in coded form both by the frequency of impulse repetition in a fiber and by the topographic relationship of retinal origin and of cortical termination. This afferent pathway does not connect retinal point to cortical point, but instead signals more synthetic geometrical arrangements, such as edges and lines in particular angular orientations (HUBEL and WIESEL, 1962, 1963; PETTIGREW, NIKARA and BISHOP, 1968a, 1968b). For example in Fig. 33A the nerve cell gives a

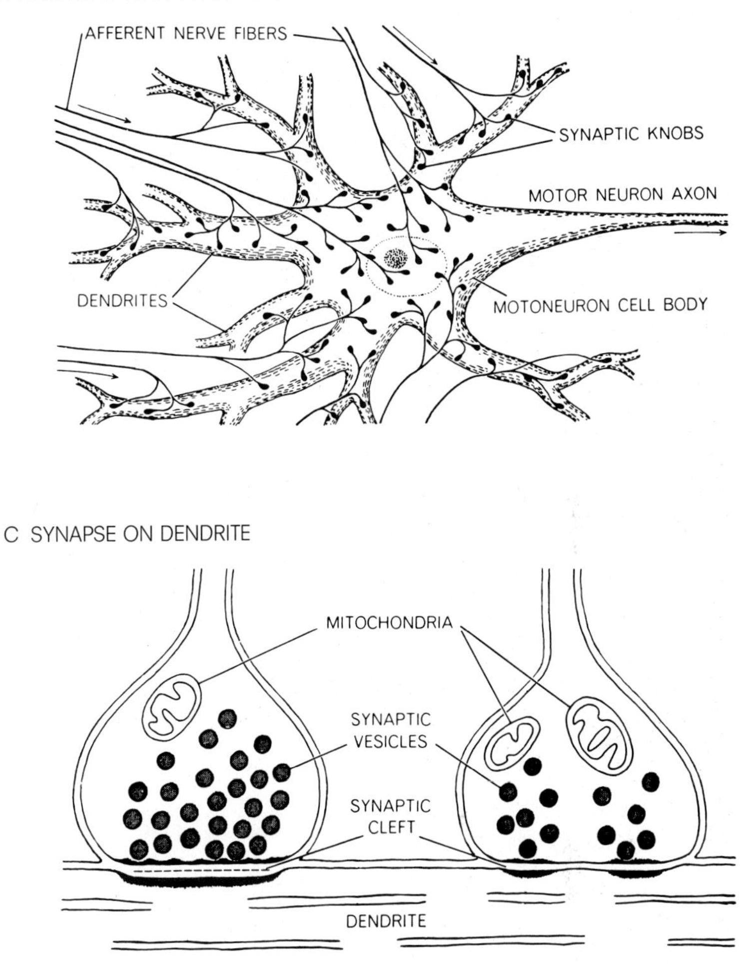

Fig. 32. B and C show neurones and synapses

burst of discharges to a bright slit that is vertical. It is not affected by a horizontal slit, and only weakly by an oblique slit. As illustrated in Fig. 33B, there is in the visual cortex a columnar arrangement of cells with similar directional sensitivities, there being cells with quite different orientations in adjacent columns. About two further stages of synthesis have been detected in the surrounds of the primary visual cortex (HUBEL and WIESEL, 1965). It can be imagined that thereafter the patterns of impulse discharge have expanded widely over the immense neuronal

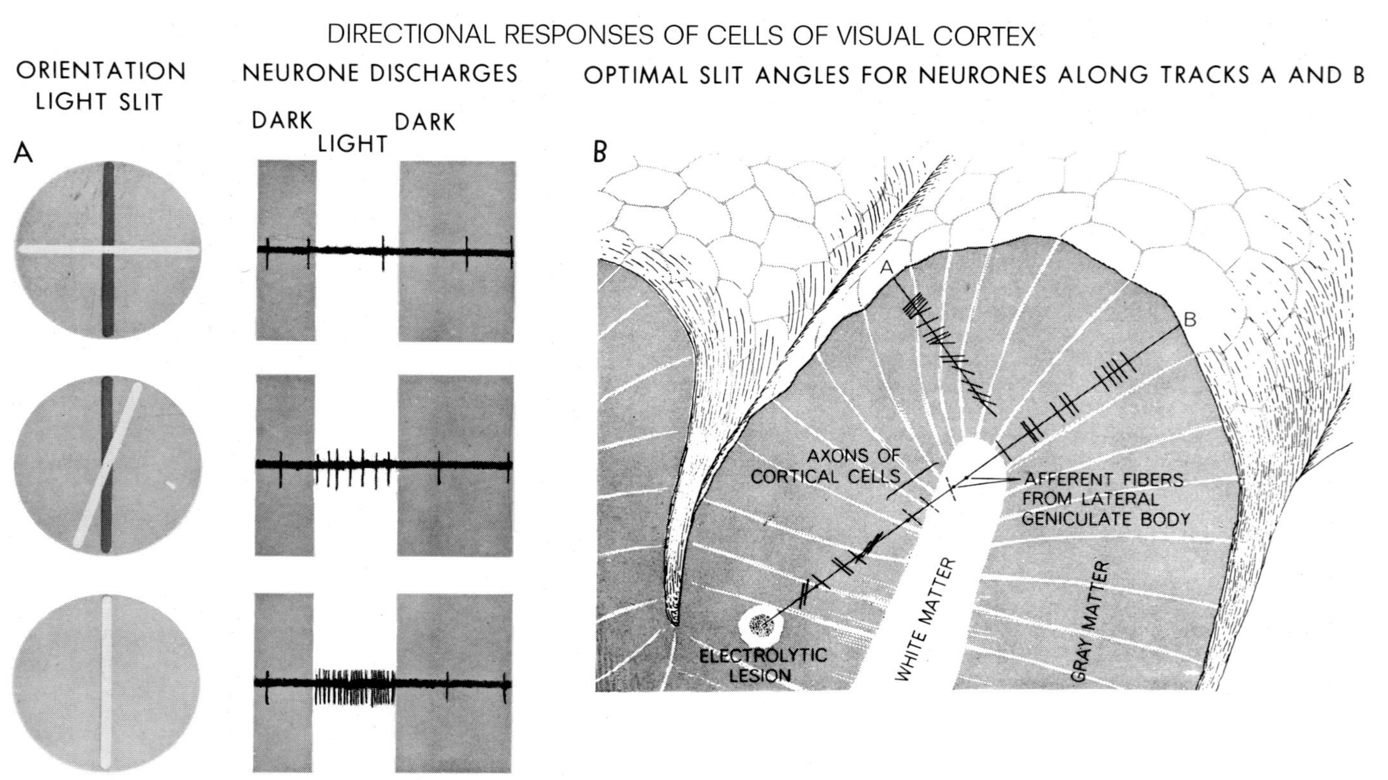

Fig. 33 A and B. Figure illustrating directional responses of cells of visual cortex. Full description in text (HUBEL, 1963)

complexity of the so-called interpretative cortex (Fig. 30), as illustrated diagrammatically in the complex patterned pathways in Fig. 12. As will be seen below, it is only at this stage of immense elaboration that there is a visual perception.

SHERRINGTON (1940) gives a most imaginative account of the problems raised by the perception of a star.

> "For instance a star which we perceive. The energy-scheme deals with it, describes the passing of radiation thence into the eye, the little light-image of it formed at the bottom of the eye, the ensuing photo-chemical action in the retina, the trains of action-potentials travelling along the nerve to the brain, the further electrical disturbance in the brain, ... But, as to our *seeing* the star it says nothing. That to our perception it is bright, has direction, has distance, that the image at the bottom of the eye-ball turns into a star overhead, a star moreover that does not move though we and our eyes as we move carry the image with us, and finally that it is the thing a star, endorsed by our cognition, about all this the energy-scheme has nothing to report. The energy-scheme deals with the star as one of the objects observable by us; as to the perceiving of it by the mind the scheme puts its finger to its lip and is silent. It may be said to bring us to the threshold of the act of perceiving, and there to bid us 'goodbye.' Its scheme seems to carry up to and through the very place and time which correlate with the mental experience, but to do so without one hint further."

Many philosophers claim that the impasse described by SHERRINGTON is illusory. They simply assert that the neuronal events are not only the necessary (as described by SHERRINGTON), but also the sufficient condition for the perceptual experience (cf. FEIGL, 1967). In other words that there is an existential inseparability of the psychic experience and the causal neuronal events, which are in the world of physics. The usual name for this physicalist hypothesis—"identity hypothesis"—is misleading, for the hypothesis states no more than in the above formulation and docs not assert that there is an identity of the psychical and the physical. It cannot be doubted that the sequence of neural events from retina to the complex patterns of neuronal discharges in the cerebral cortex carries in coded form the information needed for the visual perception, i.e. that it is the necessary condition. By asserting that it is also the sufficient condition, the identity hypothesis employs an ingenious manouvre to cover over the impasse as described by SHERRINGTON in the above quotation, but it does not solve SHERRINGTON'S problem.

Let us look again at what we can imagine going on in the neural machinery of the cerebral cortex in response to some image projected on the retina. There are firstly bursts of discharges in the simple cells as in Fig. 33A that respond to lines or edges in various specific orientations. It has to be remembered that there are at least three hundred millions of neurones in the human visual cortex (cf. Fig. 31B; SHOLL, 1956), and only a few hundreds have been experimentally investigated in mammals; nevertheless we can imagine the immense and diverse activity at this first stage of activation of the visual cortex. Within a few synaptic relays there will be activation of neurones that respond to more synthetic information—lines or slits of determinate length and width, lines or slits bent at angles, and so on (HUBEL and WIESEL, 1965). This is the limit of present investigation. Doubtless cells responding to more and more complex patterns will be discovered and it might be postulated that eventually cells will be discovered that selectively respond to abstract forms—for example to triangularity, and that this will explain our ability in the recognition of abstract forms. As expressed by SHERRINGTON (1940):

"We might imagine this principle pursued to culmination in final supreme convergence on one ultimate pontifical nerve-cell, a cell the climax of the whole system of integration. Such would be a spatial climax to a system of centralization. It would secure integration by receiving all and dispensing all as unitary arbiter of a totalitarian State. But convergence toward the brain offers in fact nothing of that kind. The brain region which we may call 'mental' is not a concentration into one cell but an enormous expansion into millions of cells. They are it is true richly interconnected. Where it is a question of 'mind' the nervous system does not integrate itself by centralization upon one pontifical cell. Rather it elaborates a million-fold democracy whose each unit is a cell."

The dynamic properties of patterned activity in tens of millions of neurones with the connectivities that have been described above defeats not only the imagination but any attempt at mathematical treatment. It must be recognized that there is no immediate experience of a perception when the visual cortex is activated. This is merely a necessary stage on the way to the much more elaborated patterned activity that is associated with consciousness. As shown by the electrical responses of the cortex, these initial stages of activation are unaltered in relatively deep anaesthesia (JASPER, 1966). Moreover with just perceptible flashes of light at least 0.2 seconds of cortical activity may be required before there is perception (CRAWFORD, 1947).

As described in Chapter V (Figs. 21, 22), the most elegant investigations on this phenomenon of perceptual delay have been carried out by LIBET and his colleagues (1966) on that part of the cerebral cortex sensitive to stimulation of the surface of the body, the somaesthetic cortex. They found that with all conditions of threshold stimulation, there was a delay of at least half a second before the onset of the experienced sensation. Evidently, there is opportunity for a great elaboration of neuronal activity in complex spatio-temporal patterns during the "incubation period" of a conscious experience at threshold level.

3. States of Consciousness

These accounts of the neural substratum of a conscious experience will serve to refute any simplistic ideas that, as soon as neural activity lights up in the cerebral cortex, there is some conscious experience, which is commonly implied in the identity hypothesis. It must first be recognized that there is intense on-going neural activity in the cortex of the awake subject in the absence of specific sensory inputs, and even in sleep there is activity (EVARTS, 1961, 1962, 1964) which is enhanced during dream periods (KLEITMAN, 1961, 1963). Then it must be recognized, as described in Chapter V, that time is required for sequential synaptic transmission and for the enormous development and elaboration of neuronal patterns before a conscious experience arises.

Finally it must be recognized that attention is required in order to register a conscious experience, and that only a very small fraction of the complex on-going patterned response of the cerebral neurones is experienced. The remainder fades unobserved because mercifully we are spared from the booming confusion that would result if we experienced in our consciousness the totality of the cerebral patterned activity at any time (MORUZZI, 1966a).

The immensity of this patterned spread throughout the neuronal pathways formed by millions of nerve cells in the brain can be imagined by thinking of an animated cartoon of finest patterned detail that is projected at a scintillating speed. As SHERRINGTON (1940) imagines, the loom weaves "a dissolving pattern, always a meaningful pattern, though never an abiding one; a shifting harmony of subpatterns." This tremendous complication of neuronal activity in my brain is required before a sensory input is perceived by me even in the rawest form; and responses involving comparison, value, judgement, correlations with remembered experiences, aesthetic evaluations, undoubtedly take much longer, with the consequence that there must be quite fantastic complexities of neuronal operation in the spatio-temporal patterns woven in the 'enchanted loom.'

When we examine the nature of all on-going perceptual experiences, it is immediately evident that it has a many-faceted nature. For example not only do we sense in our visual field the operational relations of objects but also their degree of illumination and their colour. Colour is of course especially coded information that is derived from the inputs from retinal cells that have specific photoreceptors for red, green and blue (GRANIT, 1955, 1968; RUSHTON, 1958). It is by the specificity of transmission along these coded lines that the information reaches the cortical structures where eventually there arises the perception of colour. It must be recognized that colour only appears in the picture as an experience deriving from some specifically coded patterns. There is no colour in the so-called objective world.

Equivalent statements can be made for all of our other senses. For example, in auditory perception pressure waves in the atmosphere are transduced in the inner ear, there being converted into signals of impulse discharges along the cochlear nerve fibers and so eventually to the auditory cortex where again the same immense development of dynamic pattern occurs as in the visual cortex. The perceptual experience arises, as it were, from transmutation of this pattern into our experiences of sound with pitch and loudness, melody and harmony. There is no sound with all its qualities in the external world. It is entirely our creation as specific patterns of neuronal activity in the brain are transmuted into conscious experience (cf. Chapter IV).

Perhaps when we place food in our mouths, we have the most striking example of the way in which sense organs of quite different character ultimately cooperate in giving perceptual experiences that appear to have a simple uncomplicated character. The whole range of experiences so arising are interpreted as qualities of a gustatory experience, but this is erroneous. Sensory experiences deriving from taste receptors give us only very simple reports on sweetness, sourness, bitterness and saltiness. Much more subtle additions are added by our olfactory sense, which is aroused by the movement of air with associated volatile substances to the nasal mucosa where there is excitation of the olfactory neuronal machinery that discharges impulses to the brain. In addition the placing of food in the mouth leads to a sensing from the tactile, heat, cold and pain receptors in the mouth. The pain receptors are excited by condiments and give a tang to the flavour. Our own experience immediately tells us that, when all of this various sensory input reaches the brain via quite diverse pathways, it is organized into a single gustatory experience that, at its best, delights the gourmand in us.

My conclusion from these various examples is simply that we have on the one hand all the imagined complexities of operation in the neuronal machinery of the brain which operates at levels of complexity transcending any human evaluation. And over against this there is, as defined by the problem of SHERRINGTON, the conscious experiences that are quite different in kind from any goings-on in the neuronal machinery; nevertheless the events in the neuronal machinery are a necessary condition for the experience, though in agreement with SHERRINGTON I would state that they are not a sufficient condition. Even the most complex dynamic patterns played out in the neuronal machinery of the cerebral cortex are in the matter-energy world. Transcending this level, and in emergent relationship from it, is the world of conscious experience—called the "Noosphere" by TEILHARD DE CHARDIN (1959).

This belief in the existence of a distinct world of states of consciousness was expressed in my three earlier quotations from POPPER, WIGNER and SCHRÖDINGER, but of course it has had a long history, and in particular is associated with the great contribution to philosophy by DESCARTES. It was he who first recognized that perceptions only occur when the signals from the peripheral sense organs are transmitted via the nerves to the brain and that the goings-on in the brain give the conscious experience. Reciprocally this conscious world of DESCARTES also can exert an action out to the external world by means of the brain and thence via nerves to muscles. This is the pathway postulated for free-will, as described in Chapter VIII.

It has been stated there that freedom of the will is a primary fact of experience, and that the formulation of the problem arising from this experience should be the inverse of its usual statement. The problem is to discover in the brain the functional properties that give it the requisite responsiveness, so that, when I consciously will an action, I call forth responses that lead to the desired muscular movements. To define the problem more precisely, it must be postulated that goings-on in my consciousness, such as dispositional intentions, are able to effect changes in the patterns of neuronal activities in my brain that eventually result in modifications of the discharges down the pyramidal tract to motoneurones and so to muscles.

So far, there have been merely initial, tentative probings in relation to this hitherto intractable problem of free-will, which is the reciprocal of SCHRÖDINGER'S problem. It has been argued in Chapter VIII that both physics and physiology are too primitive to allow even the proper formulation of the problem, let alone its solution. One can surmise from the extreme complexity and refinement of its organization that there must be an unimaginable richness of properties in the active cerebral cortex, giving it the property of being a "detector" in liaison with dispositional intentions in the world of conscious experience, as I postulated many years ago (ECCLES, 1953). Of course I recognize that by far the greater part of our behaviour is not controlled by conscious decisions "freely made." There are all levels of control from deliberate decisions, to the semi-automatic with operation of sub-routines of behaviour, to fully automatic routines; nevertheless from moment to moment we can switch our behaviour pattern from being routine to being consciously controlled, as for example occurs with an experienced car driver when confronted by an emergency. An even more acceptable example of transformation from an automatic routine to conscious control occurs in knitting when a stitch is dropped!

4. The Three-World Concept of Popper

There have been many philosophical developments related to the two-world postulate of DESCARTES, that is generally called dualism. I myself until recently believed that this two-world concept provided an adequate explanation of all of our experience and knowledge. However, in two recent publications Popper (1968 a, 1968 b) has developed in a remarkable manner his concepts of three worlds, and I now wish to present this important new development with some modifications that I feel are desirable in the light of our knowledge of neurophysiology.

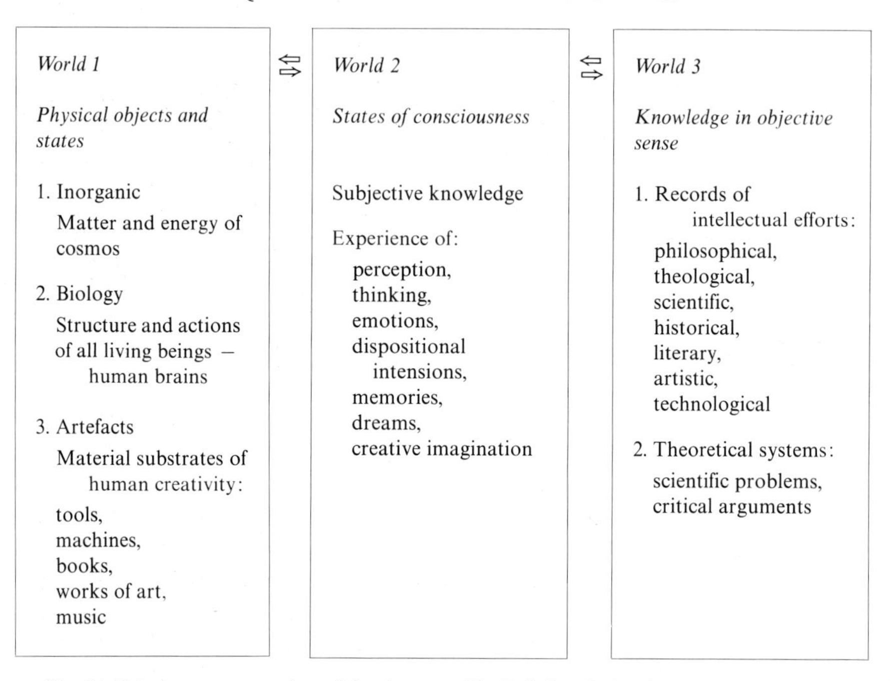

Fig. 34. Tabular representation of the three worlds. Full description in text

Fig. 34 represents an attempt to give a diagrammatic expression of the postulated contents of the three worlds. World 1 is the world of physical objects and states and as such comprises not only the inorganic matter and energy of the cosmos, but also all biology—the structures and actions of all living beings, plants and animals and even human brains. It also comprises the material substratum of all man-made objects or artefacts—machines, books, works of art, films and computers.

World 2 is the world of states of consciousness or mental states. For present purposes we need not raise the issue of animal consciousness, about which we can be agnostic. World 2 is the world that each of us knows at first hand only for oneself, and in others by inference. It is the world of knowledge in a subjective sense, and comprises the ongoing experiences of perception, of thinking, of emotions, of imaginings, of dispositional intentions and of memories.

By contrast, World 3 is the world of knowledge in the objective sense, and as such has an extremely wide range of contents. In Fig. 34, there is an abbreviated list. For example it comprises the expressions of scientific, literary and artistic thoughts that have been preserved in codified form in libraries, in museums and in all records of human culture. In their material composition of paper and ink, books are in World 1, but the codified knowledge conveyed in the print is in World 3, and similarly for

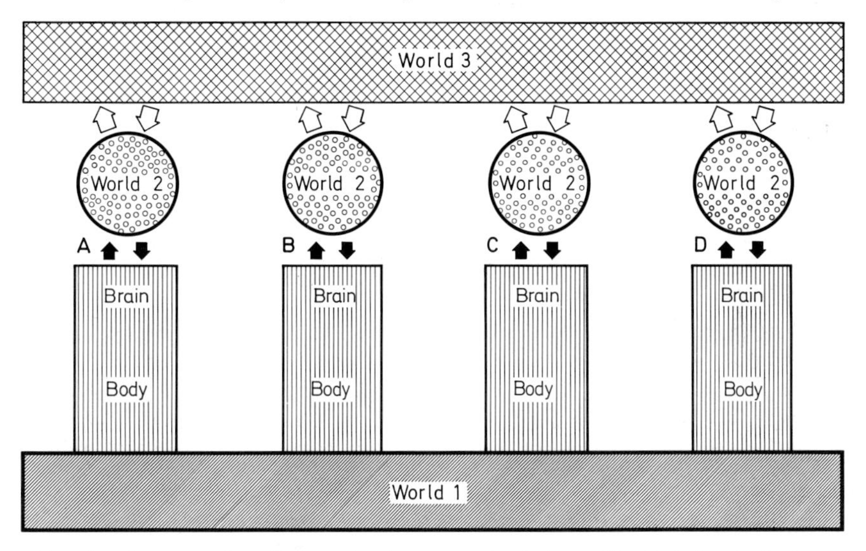

Fig. 35. Information-flow diagram showing interaction between Worlds 1, 2 and 3 for four individuals. World 3 is represented as being "impressed" on a special part of World 1 (books, pictures, films, tapes, etc.) that forms an indefinitely extending layer. It is to be understood that any individual is able to move into relationship with any particular part of World 3 that is of interest to him. Further description in text

pictures and all other artefacts. Most important components of World 3 are the theoretical systems comprising scientific problems and the critical arguments generated by discussions of these problems. In summary it can be stated that World 3 comprises the records of the intellectual efforts of all mankind through all ages up to the present—what we may call the cultural heritage.

Fig. 34 shows diagrammatically the manner of the interactions that Popper specifies for these three worlds, namely that there is reciprocal transmission between 1 and 2 and between 2 and 3, but that 1 and 3 can interact only by mediation of World 2. A further development of the 3-world concept is displayed in Fig. 35, where the bodies and brains of 4 humans, A, B, C, D, are shown projecting upwards from the general mass of World 1, but still being in World 1 as is signified by the linear shading. Each human is shown to be in reciprocal relationship to a unique World 2 which is his world of knowledge in the subjective sense, and hence private to him. However it is in a state of continual reception and transmission to and from World 1, and so with other humans, by mediation of the associated body and brain. As ordinarily envisaged, extrasensory perception would entail direct transmission between the subjective states of the World 2's of two humans (cf. HARDY, 1965). Finally Fig. 35 shows diagrammatically and symbolically the reciprocal communication that is postulated by POPPER between the individual Worlds 2 and World 3, the world of the objective spirit, that potentially can be shared between any number of humans. It must be emphasized that Fig. 35 is an information-flow diagram, and *not* a topographical diagram displaying spatial relations. In the intellectual and creative life of the spirit it can be envisaged that there is an intense on-going traffic in both directions in the continual critical wrestling with problems of understanding and expression. Thanks to the sum total of creative intellectual efforts of mankind World 3 is immensely rich and extensive and during a whole life-time it is not possible for one individual to do more than sample a minute fraction, and only rarely can a significant addition be made. The central task of the humanities is to understand men, while that of the natural sciences is to understand nature. Since in both cases this understanding achieves expression in language, both can be regarded as branches of literature, and both have an honoured place in World 3.

It will be appreciated that in Figs. 34 and 35 the lines of communication are merely symbolic and are far removed from the actual modes of communication as defined neurophysiologically. Fig. 36 represents an attempt to give in outline the principal pathways for the flow of information for a single human, but even at this minimal level of symbolic representation there has been a great complication of the simplicity of Fig. 35.

The body has been separated from the rest of World 1 that is depicted as a small basal fragment, and an arrow shows the line of reception from World 1, via receptor organs and the afferent pathways (Aff. path) up to the brain. Reciprocally there are shown descending neural pathways (Eff. path) leading from the brain to effector organs such as muscle that in turn act on World 1 as shown by the arrow. A transverse arrow in the brain depicts the various subconscious reflex actions, and the vertical arrows continue upwards to the upper rectangle which represents that part of the brain whose activities are in liaison with the conscious mind of World 2, as is shown by the further upwardly directed arrows. This sequence of upward arrows represents the single arrows shown in Fig. 35 from World 1 to World 2, but it will be recognized that all of the brain is in World 1. It acts merely as a selective channel for the flow of information. Similarly the downward arrows in Fig. 36 from World 2 to World 1 represent the single downward arrow in Fig. 35, and depict the flow of

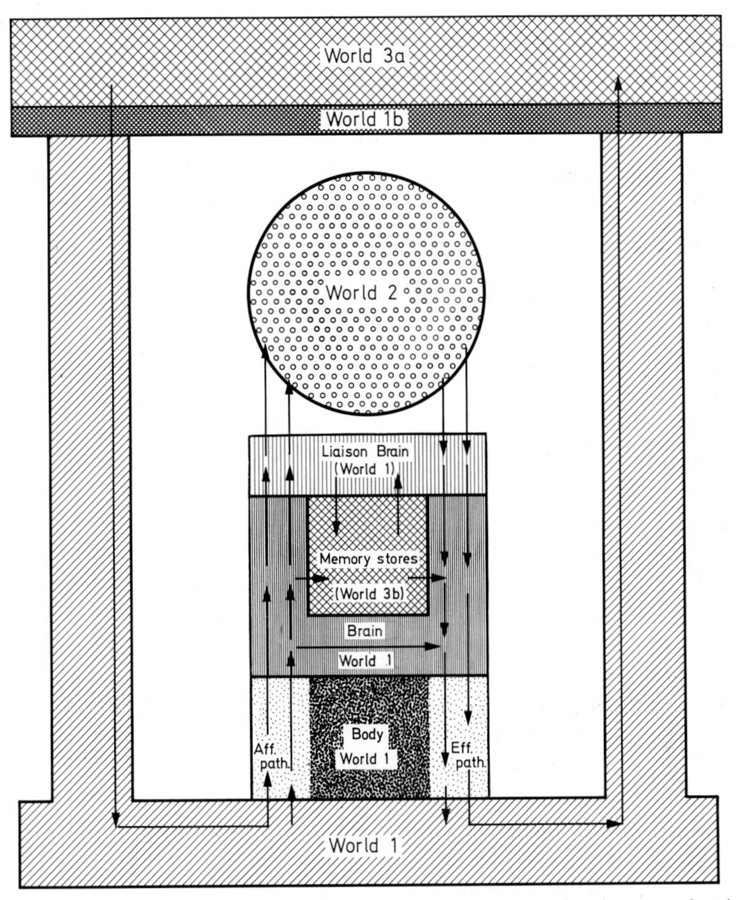

Fig. 36. Information-flow diagram representing modes of interaction between the three Worlds as shown by the pathways represented by arrows. It is to be noted that, except for the liaison between the brain and World 2, all of information occurs in the matter-energy system of World 1. For example, in the reading of a book, communication between the book and the receptor mechanisms of the eye is subserved by radiation in the band of visual wave lengths. As in Fig. 35 it is to be understood that any individual can at will range widely in his relationship to World 3. For further description see text

information whereby a conscious decision in World 2 achieves expression in World 1 by means of a muscular movement, which is represented by the lowest arrow (Eff. path).

The diagram of Fig. 36 further recognizes the fact that receptor organs and their associated neural pathways provide the only channels of communication by which the brain can "sense" events in the external world, which is of course World 1 and not World 3. World 3 cannot be directly sensed, but is shown encoded on a special part of World 1, World 1b, which comprises the paper and ink of books for example. The printed patterns of the letters are transmitted in the various neural codes from the retina to the visual cortex (cf. Fig. 31 A) and eventually to the decoding centers of the brain, the so-called interpretive cortex (cf. Fig. 30), that in turn transmits as liaison brain in Fig. 36 to give the perceptual experiences of World 2. Reciprocally the pathway from World 2 to World 3, as for example in the expression of a scientific or artistic idea, goes, as shown by the arrows (Eff. path), to effector organs and thence to a special part of World 1 (I b) on which the encoded message is impressed. Quite complex pathways have replaced the simple arrows reciprocally connecting Worlds 2 and 3 in Fig. 35.

It has seemed necessary to introduce one further complication in the brain in Fig. 36, and that is the central square which represents the memory stores of components that can I think properly be regarded as belonging to World 3 (World 3b). For example much that is destined for eventual impression in the appropriate coding in World 3 may be held in such memory stores and is immediately available in discussions on scientific, artistic or other cultural matters. The largest contributions to a scientific symposium usually arise in this way from the memory stores of the participants rather than from any manuscripts they may bring along. Another example would be the Homeric poems or the Icelandic Sagas that survived for centuries in memory stores of successive generations of bards. Arrows entering and leaving this central memory store depict the flow of information inwards for storage or outwards on retrieval.

In order to illustrate the independent existence of the third world, POPPER (1968a) considers two "thought experiments":

"Experiment (1). All our machines and tools are destroyed, also all our subjective learning, including our subjective knowledge of machines and tools, and how to use them. But *libraries and our capacity to learn from them survive*. Clearly, after much suffering, our world may get going again.

Experiment (2). As before, machines and tools are destroyed, and our subjective learning, including our subjective knowledge of machines and tools, and how to use them. But this time, *all libraries are destroyed also*, so that our capacity to learn from books becomes useless.

If you think about these two experiments, the reality, significance and the degree of autonomy of the third world (as well as its effects on the second and first worlds) may perhaps become a little clearer to you. For in the second case there will be no re-emergence of our civilization for many millennia."

If we think of the history of the human race, it will be appreciated that in the case of Experiment (2) man is transported far back into prehistory and would have to begin the long climb up that characterized the tens of thousands of years from Neanderthal man through to the

Cro-Magnon man and so eventually through the historical epochs. On the other hand even the great destruction of the Roman Empire was incomplete, there being islands of culture with written records that survived and eventually the recovery of classical culture came in hundreds of years, largely through the isolated monastic centers and the Arabian scholars.

It is important to recognize that knowledge in the objective sense (World 3) is a product of human intellectual activity, and, although it has an existence independent of a knowing subject, it must be potentially capable of being known, that is, it is largely autonomous, but not absolutely and finally autonomous. I can give as an example the Linear B scripts of the Minoan civilization which have only recently been deciphered by MICHAEL VENTRIS. The information symbolically carried by the scripts was nevertheless in World 3 because they were potentially capable of being deciphered.

POPPER (1968b) goes on to suggest that the three worlds have an objective reality and

> "that a subjective mental world of personal experiences exists (a thesis denied by the behaviourists) ... that it is one of the main functions of the second world to grasp the objects of the third world. This is something we all do: it is part of being human to learn a language and this means, essentially, to learn to grasp *objective thought contents* ... that one day we will have to revolutionize psychology by looking at the human mind as an organ for interacting with the objects of the third world; for understanding them, contributing to them, participating in them; and for bringing them to bear on the first world."

The third world has been called by POPPER the world of the objective spirit. It may seem like Plato's world of forms and ideas, but in reality it is quite different. For Plato the third world comprised eternal verities that provided ultimate explanations and meaning for all of our experience, and our efforts were concerned in trying to grasp and understand these eternal verities. They were to be gazed at and intuited, but not criticized or changed. They were a third world of possible objects of thought which were transcendental. It is clear that this Platonic world is different from POPPER's third world, which is man-made and arises from our efforts to understand and make intelligible World 1 and even World 2. We do this by formulating problems, and wrestling with problems in our attempts to derive deeper and deeper understandings of the whole of our experience. This is a shared enterprise of creative and critical mental effort. In this enterprise we have the interactions between World 1 and the experiencing self of World 2 on the one hand, and on the other hand the interaction between the experiencing self (World 2) and the world of the objective spirit with its ideas in the objective sense (World 3). This is a world of possible objects of thought, which is the world of civilization and of culture from all ages to the on-going creativity today.

5. The World of States of Consciousness

As we have seen above (Chapters IV, V and VI) the unimaginable organized complexity of the cerebrum has caused the emergence of properties (World 2) which are of a different kind from anything as yet related to the matter of World 1 with its properties as defined in physics and chemistry.

World 2 is defined as the world of subjective experience, and hence is private to each individual; nevertheless communication by language establishes that other individuals have comparable subjective experiences. For this purpose the lower functions of language that man shares with some animals are useless, that is expressive cries and stimulative signals (POPPER, 1962). The higher functions of language, called the descriptive and argumentative (POPPER, 1962), or the propositional (ADLER, 1967) are essential for this communication. After meticulously reasoned arguments ADLER (1967) comes to the conclusions that man alone has a propositional language and that this language can be employed only by subjects who have conceptual thought, which is essentially thought related to the components of World 3. This thought transcends the perceptual present. It concerns the uniquely human development, utilizing concepts and symbols and rational arguments. By contrast the behaviour of animals is derived from their perceptual present and their background conditioning. When confronted by a situation, they rely on trial and error rather than on the attempt to understand and to act rationally. They also rely on imitative behaviour.

In his book "An Essay on Man" CASSIRER (1944) states that:

> "without symbolism, the life of man would be like that of the prisoners in the cave of Plato's famous simile. Man's life would be confined within the limits of his biological needs and his practical interests; it could find no access to the ideal world which is opened to him from different sides by religion, art, philosophy, science."

This ideal world will be recognized as the world of the objective spirit – the third World of POPPER. There is no evidence that animals share this World even in the smallest degree. In this fundamental respect men are radically different in kind from animals.

I find myself in complete agreement with ADLER'S (1967) expression of the human situation.

> "On this interpretation of the observed fact that linguistic animals differ in kind from non-linguistic animals, is man a person rather than a thing? The answer is affirmative if the line that divides persons from things can be drawn by such criteria as conversational ability, the ability to engage in meaningful discourse, and the ability to give and receive reasons or arguments. By these criteria, men are at present the only beings on earth that are persons. All other animals and machines are things—at least in the light of available evidence. The special worth or dignity that belongs exclusively to persons, the respect that must be accorded only to persons, the fundamental imperative that commands us to treat

persons as ends, never solely as means—all these are thought to obtain on this theory of what is involved in being a persons."

DESCARTES (Trans. 1931) recognized the definitive nature of the linguistic test:

"If there were machines which bore a resemblance to our body, and imitated our actions so far as it was morally (i.e., practically) possible to do so, we should always have two very certain tests by which to recognize that, for all that, they were not real men. The first is that they could never use speech or other signs as we do when placing our thought on record for the benefit of others. For we can easily understand a machine's being constituted so that it can utter words, and even emit some responses to action on it of a corporeal kind, which brings about a change in its organs; for instance, if it is touched in a particular part, it may ask what we wish to say to it; if in another part, it may exclaim that it is being hurt and so on. But it (could) never happen that it (would) arrange its speech in various ways, in order to reply appropriately to everything that may be said in its presence, as even the lowest type of man can do .... This does not merely show that the brutes have less reason than men, but that they have none at all, since it is clear that very little is required in order to be able to talk."

This conversational test provides the rationale for the fanciful proposal by TURING (1950), namely that, if a robot, a so-called Turing machine, could be constructed that passed the conversational test, then it would be established that complex material structures are adequate to explain all of man's experiences and performances, which would be a triumph for the identity hypothesis. I do not believe that such a robot will ever be constructed. I am appalled by the naïveté of the statements and arguments that are made by the proponents of the computer simulation of man. I am in complete agreement with the extremely well documented and destructive criticisms of JAKI (1969). Much more is involved in the human performance than being adept in conversation, or successful in checkers or chess. At present computers are being used to display that *with suitable programming* they have expertise in simple games like checkers, but only moderate performance in complex games like chess (HUSTON SMITH, 1967). The expenditure of thirty million dollars has revealed virtually insuperable problems in mechanical language translation.

I now return to the four initial quotations by SHERRINGTON, POPPER, WIGNER and SCHRÖDINGER. The discussions stemming therefrom have amplified and corroborated the belief that central to each human being is the primary reality of conscious experience in all the richness and diversity that characterizes World 2 existence. Furthermore this experience is self-reflective in the sense that we know that we can know. Our ultimate efforts are to understand this primary reality in relation to the secondary realities of the matter-energy world (World 1) and of the world of objective thought that embraces the whole of civilization and culture (World 3). We as experiencing beings must be central to the explanations, because all the experiences derived from Worlds 1 and 3 are recognizably dependent on the manner in which we obtain information by means of the transductions effected by sense organs and the coded transmission to our brains.

Scientific investigations utilize technical procedures involving the utmost subtlety and power of resolution, and there may be the most sophisticated processing of data by computers; nevertheless eventually it has to be sensed by the receptor organs of the scientist and transmitted to his brain, there to undergo most elaborate developments into dynamic patterns of the neuronal machinery that by transmutation give him perceptual experiences in his World 2. As WIGNER (1964) states:

> "the measurement is not completed until its result enters our consciousness. This last step occurs when a correlation is established between the state of the last measuring apparatus and something which directly affects our consciousness. The last step is, at the present state of our knowledge, shrouded in mystery and no explanation has been given for it so far in terms of quantum mechanics, or in terms of any other theory."

In this mysterious way we derive the raw data of a scientific experiment, which is merely the beginning of the scientific process. There then ensues the transactions between the World 2 of the scientist and World 3, which is the world of problems, arguments, hypotheses, where critical evaluation and creative imagination are supreme. This is the arena in which scientific battles are fought, firstly by the scientist alone as he evaluates his hypotheses against the experimental results, and secondly, if there is eventual publication, in the tough critical world (World 3) of scientific disputation (Chapter IX).

6. Self Awareness and Death Awareness

This whole cultural life of World 3 is exclusively human, and related to it is the unique self-consciousness that each of us experiences and that we can discover in other human beings through communication by propositional speech as DOBZHANSKY (1967) comments:

> "Self-awareness is, then, one of the fundamental, possibly the most fundamental, characteristic of the human species. This characteristic is an evolutionary novelty; the biological species from which mankind has descended had only rudiments of self-awareness, or perhaps lacked it altogether. The self-awareness has, however, brought in its train somber companions—fear, anxiety, and death-awareness."

DOBZHANSKY (1967, 1969) goes on to develop the thesis that the death-awareness attending on self-consciousness can be recognized as being experienced by those primitive men that practised ceremonial burial customs. The most ancient known burials are Neanderthal man in Palestine about 100,000 years ago. We can speak of man's ancestors as by that time being at the dawn of humanity and having become self-conscious beings. We know that their brains were not inferior to ours in size (cf. Chapter VI).

It is widely held amongst scientists that everything is reducible to physics and chemistry. For example that thoughts are nothing but on-going activities in the neuronal machinery of the brain (physicalism; cf. FEIGL, 1967) and that the actions of all living organisms including human brains are nothing but physics and chemistry (cf. YOUNG, 1951b; CRICK, 1966). We owe to POLANYI (1966, 1967a, 1968b) in particular a critical examination of this "nothing-buttery" reductionist dogma. He develops the concepts of hierarchical levels stemming from the world of the fundamental particles of physics through atoms and molecules to machines and living organisms. At each level special boundary conditions provide the operational principles that utilize the properties of the lower level to give new emergent properties. For example a machine such as a watch obeys the laws of physics and chemistry, but is designed with operational principles for a specific purpose. Similarly, living organisms obey the laws of physics and chemistry, but with the shaping given by the framework of the living structure with its genetically controlled development. Evolution is the amazing biological process by which living organisms have developed structures and operational principles highly adapted for survival, replication and multiplication (cf. Chapters IV and VI).

It is my contention that, just as in biology there are new emergent properties of matter, so at the extreme level of organized complexity of the cerebral cortex, there arises still further emergence, namely the property of being associated with conscious experiences. In some such manner we may eventually account for the emergence of self-awareness in Neanderthal man – and in all subsequent men (Chapter VI). For the present we may regard the cerebral cortex as providing the necessary condition for self-awareness. We now have to consider if it also provides the sufficient condition.

7. The Concept of the Soul

I have been long in making direct reference to the soul that forms the second component of my discourse. I have deliberately refrained because I am talking to a modern sophisticated audience that presumably has been conditioned to be extremely suspicious of such a theological term! Moreover even in theological discussions the connotation of the word "soul" varies widely. I have now presented a scientific-philosophical basis on which to develop what I believe to be the only tenable position with respect to the brain and the soul. I submit that the Aristotelian-Thomist view that "the soul is the form of the body" is no longer tenable, as was realized already by DESCARTES. However in the historical surveys of the theological and philosophical disputations that are given by VON HÜGEL (1912), BAILLIE (1934), MACNEILLE DIXON (1937), PELIKAN (1962), to name but a few, I find compatibility with the suggestion that the subjective component of each of us in World 2, the conscious self, may be identified as the soul.

The component of our existence in World 2 is non-material and hence is not subject in death to the disintegration that affects all components of the individual in World 1—both the body and the brain, though of course it is deprived of all communication with World 1 and World 3, and hence, for each of us, all experience as we know it must cease in oblivion. But we may ask in hope: Need that oblivion be unending? It will be recognized that this identification of World 2 as the soul is essentially the Cartesian position, but without commitment to any explanation of how the soul is "attached" to the brain. A similar position was adopted by SHERRINGTON (1940) in his identification of the self or mind-concept with the soul-concept, though he went on to state:

> "When on the other hand the mind-concept is so applied as to insert into the human individual an immortal soul, again a trespass is committed. The very concomitance of the two concepts, which seems a basal condition of our knowledge of them, is thrown aside as if forgotten. Such amplification of the one concept may be legitimate for a revealed religion .... But as an assertion on the plane of Natural Knowledge it is an irrational blow at ... that very harmony which unites the concepts as sister concepts. It severs them and drives one of them off, lonelily enough, on a flight into the rainbow's end."

However many years later (Feb. 24, 1952) he told me with great passion, "For me now the only reality is the human soul." He made this statement during a deeply moving discourse and I did not break in to ask if this statement was an act of faith expressing a religious conviction, though I thought he so implied. Nine days later he was dead.

The story of man's thoughts on the meaning of life and on the ultimate human destiny in death has been collected in two books by CHORON (1963, 1964). Myths and religions and philosophies have been concerned with this tragic enigma of "ultimate concern" that faces each one of us. Is human destiny but an episode between two oblivions? Or can we have hope that there is meaning and transcendent significance in the wonderful, rich and vivid conscious experience that is our birthright?

And that brings me to assert that any fundamental question in philosophy must be considered in the full context of related questions, and never in some arbitrary isolation. The question of death-awareness and self-annihilation must not be discussed except in relation to the question of birth and the subsequent self-actualization, which has been expressed by Plato in the Phaedo. As I have stated above and in Chapters IV, V, VI, and as I have argued previously (ECCLES, 1965a, 1967, 1969a), I believe that my experiencing self is only in part explained by the evolutionary origin of my body and brain, that is of my World 1 component. It is a necessary but not a sufficient condition. About the origin of our world of conscious experience (World 2) we know only that it can be described as having an emergent relation to the evolutionary development of the human brain. The uniqueness of the individual that I experience myself to have cannot be attributed to the uniqueness of my DNA inheritance, as I have already argued (Chapter V; ECCLES, 1965a). Our coming-to-be is as mysterious as our ceasing-to-be at death. Can we therefore not derive hope because our ignorance about our origin matches our ignorance about our destiny? Cannot life be lived as a challenging and wonderful adventure that has meaning to be discovered?

It has been argued above that man differs *radically in kind* from other animals. As a transcendence in the evolutionary process there appeared an animal radically differing from other animals because he had attained to propositional speech, abstract thought and self-consciousness, which are all signs that a being of transcendent novelty had appeared in the world—creatures existing not only in World 1 but realizing their existence in the world of self-awareness (World 2) and so having in the religious concept, souls. And soon these humans began utilizing their World 2 experiences to create another world, the third World of the objective spirit. This World 3 provides the means whereby man's creative efforts live on as a heritage for all future men, so building the magnificent cultures and civilizations recorded in human history. Do not the mystery and the wonder of this story of our origin and nature surpass the myths whereby man in the past has attempted to explain his origin and destiny?

Are we to build further in this record of man's greatness in World 3? For that we require a free society, for only under such conditions can there be a free-flowering of man's creative powers in the sciences and the humanities (Chapter VIII and IX). For those who have eyes to see there is the tragic alternative of a totalitarian aggressor whose power structure is on the one hand the might and threat of armaments and on the other an enslaved population. This totalitarian system is based on a false materialist philosophy of man, which denigrates him to be merely a clever animal, and hence a thing. If we do not recognize this terrible threat, if we lose our nerve, then human freedom will be eliminated as effectively as in ORWELL'S 1984 (ORWELL, 1949). Despite the technology that would survive in a totalitarian world, enslaved mankind would have lost its soul in a long dark night of cultural and intellectual barbarism. In this tragic hour we must know what we are fighting for, we must appreciate man's greatness, we must regain our faith and hope in man and his destiny—else all is lost.

example I can instance the whole effort of archaeology that can be considered as an attempt to uncover and discover the World 3 of ancient civilizations. Archaeologists are particularly sensitized to recognize the most primitive examples of human artefacts and to distinguish them from objects that were formed entirely by "natural" causes. As examples of more sophisticated archaeology, we can instance the efforts to decode written languages. Until this has been accomplished, the presumed symbolic representation of a language is not in World 3 for all practical purposes, though potentially it is part of World 3. A recent example is cited in Chapter X in the deciphering the linear B scripts of Minoan civilization.

Having defined the status of World 3, we can now come to the interrelations of Worlds 1, 2 and 3. The interactions of World 1 with 2 and 2 with 1 are the familiar processes of perception on the one hand and willed action on the other. We can regard these interactions as the continual and intense ongoing activity that is characteristic of conscious life. In World 2 we put all the subjective states that are directly known to us. I would relegate to World 1 all subconscious reactivity, even that requiring high levels of brain action, for example the subconscious mind postulated by psychiatrists. There seems no reason to postulate that such subconscious activities involve more than the ongoing events studied by the physiologists and biochemists of the higher nervous system.

These considerations bring us to the statement that World 2 includes the whole world of conscious experience and nothing else. That is, it is the world of primary reality defined by EUGENE WIGNER (1964). As stated above, particularly in Chapters IV, V and X, it includes the whole of conscious perception, action, memory, creative thought and emotions. It is important to recognize that conscious perception does not merely give information about the external world, but also about the happenings within the body such as are signalled by perceptions of pain, hunger, thirst and all the much more subtle sensing that occurs with perceptions of emotional states—joy, sadness, fear, despair, etc., and also in our experiences of beauty, wonder, awe, etc.

It is of the greatest significance for education to consider the interactions of Worlds 2 and 3. In World 3 we have the total assemblage of culture in all aspects. Of course as educated men we each carry in our subjective World 2 much of the culture that we have stored in memory and that in Fig. 36 was located in the brain and labelled World 3 b. In fact almost all of culture was so carried before there were the technical developments of coding knowledge in some enduring manner, as for example in the writing of a language and in other symbolic expressions in art and music. In recent times we have added significantly to these modes of coding by the use of films and microfilms, recording tapes, punch cards and computer memories. In his discussion of the interaction of Worlds 2 and 3 POPPER assumes that this interaction occurs directly. For example from World 2 to World 3 we would have ongoing exchanges involving theories, problems and discussions and in the reverse direction cultural, scientific and artistic judgements. I can agree with POPPER'S statement that Worlds 1 and 3 do not directly interact, but only through the mediation of the second world – the world of subjective or personal experience. However a neurophysiologist would recognize that World 3 as such cannot be directly sensed by World 2. As physiologists we have to insist that the coded knowledge of World 3 has to be transmitted by World 1 mechanisms, such as is involved in giving the subject visually sensed data (reading) of the coded information inscribed upon World 1 objects (books). There is then transmission through the visual mechanisms of the human eye and brain (World 1) eventually to appear as the dynamic ongoing patterned activity of tens of millions of neurones in the cerebral cortex (cf. Fig. 33) – which is still World 1. This pathway is shown diagrammatically in Fig. 36. At this stage the neurophysiologist can say no more. In a way not at all understood this coded pattern of circulating nerve impulses in the so-called liaison brain appears as a conscious experience in World 2 of the observer. It is not at all possible to develop further the physiological concepts related to this World 2 and 3 interaction in the manner essayed in Chapter VIII. All that need be said is that it is immensely more complex than appears from POPPER'S formulation of his interactions of the Worlds 2 and 3 – and far beyond anybody's comprehension even at the vaguest level of understanding.

2. Education and the Third World

Education can be thought of as the training that gives the ability of the conscious selves of the second world to grasp and understand the third world. It is something that we all have learned to do. For example, it is part of being human to learn a language and this means essentially to learn to grasp objective thought contents, which of course are in the third world, and which contain in essence all of the science, technology and the creative arts such as literature and music. The third world is in a sense like Plato's world of forms and ideas, but in reality it is quite different because it is man-made and does not belong to some world of eternal verities such as Plato conjectured. In fact it has been the task and performance of man the creator to have developed the objective contents of thought in the third world in the magnificent manner that gives us all of our civilization. It has been a shared enterprise of creative and critical mental effort, i. e. the creative activities of the second worlds of countless individuals, and it is an on-going creative process. A civilization flourishes when its intellectual leaders, the *Homo individuales* of FONTENAY (1968), further develop and refine the objective contents of thought of the third world, utilizing for this purpose all the appropriate records of the past and remoulding them in the light of their own creative insights.

The third world as defined by POPPER is a product of human endeavour, though in its ontological status, it is autonomous. The third world has grown far beyond the grasp and appreciation of any one man. Its growth is largely due to a positive feed-back effect which derives from the challenge of autonomous problems, many of which may never be mastered. In addition there will always be the exciting experience of discovering new problems, which of course has been the great success of science in these last decades. Education in its highest form can be considered as the effort to give to the subjective minds in World 2 a deeper and broader understanding of the objective thought that men have developed in World 3. The greatest and most important effort of education must be to inform and inspire each generation by providing them both with access to the objective knowledge of World 3 and with critical judgement in relationship to this world (ECCLES, 1966e).

Education is an on-going process throughout the whole adventure of life—that is for those that take life as an adventure. It can lead to an enrichment of experience and to the thrill of deeper understandings. As POPPER (1968a) states:

"This self-transcendence is the most striking and important fact of all life and all evolution, and especially of human evolution .... This is how we lift ourselves by our bootstraps out of the morass of our ignorance; how we throw a rope into the air and then swarm up it—if it gets any purchase, however precarious, on any little twig."

As against this program of illimitable illumination we have in this present world many developing threats. There is first of all a cynical materialism and irrationalism and a growing feeling of the hopelessness and meaninglessness of life. But even more meancing is the totalitarian tyranny repressing the free expression and the creativity of man. This tyranny operates by the distortion of truth and the subversion of values so that words are no longer employed to convey meaning and truth but instead are used as weapons. I can quote from a brief article by HERMAN WOUK in relationship to the Pueblo disaster and the dilemma confronting Commander BUCHER.

"The lack of communication between ourselves and the Communists is radical. It strikes down to the nature of man and the use of words. In the West, individual life has high value, and words are tested against a measuring-rod called truth. In the realm of Communism, the individual loses himself in the state. Words are only tools for politics and war. The lie as such does not exist; truth is relative to state needs. Trapped in that other world, Commander BUCHER signed a document of tool-words, to keep alive the Americans he had saved with his first decision. Now, back in his own world, he must answer for the fact that in our terms the words were lies."

One is confronted by a future that could be a catastrophic fall into the tragic pit of the DANTEAN hell depicted in GEORGE ORWELL'S 1984. We have the alternative that we must be inspired to arise from our present troubles and confusions and continue with the great civilizing and cultural achievements that are the glorious story of mankind. There have of course been great set-backs and long periods of darkness, but again and again the spirit of man has triumphed and created new levels of existence in the objective world of the spirit. POPPER (1968a) states:

"What may be called the second world—the world of the mind—becomes, on the human level, more and more the link between the first and the third world: all our actions in the first world are influenced by our second world grasp of the third world.

"We do not mould or 'instruct' this world (of objective knowledge) by expressing in it the state of our mind; nor does it instruct us: both, we ourselves and the third world grow through mutual struggle and selection. This, it seems, holds at the level of the enzyme and the gene—the genetic code may be conjectured to operate by selection or rejection rather than by instruction or command—and through all levels, up to the articulate and critical language of our theories."

As I can vividly remember from many of my life experiences, a young student is confronted by the tremendous unknowns of World 3, as for example exhibited in a great library of books. He tends to shrink back from it as being forever beyond his limited experience. He naturally feels overawed by this incredible assemblage of the literature embodying World 3 knowledge from all ages. One has the experience of the intimidating extent of knowledge extending apparently to an infinite horizon. How can one learn to adventure for oneself over fields developed by the great creative minds of the past? This brings me to the initial role of the University teacher, who should be a guide to exploration so that the student eventually will learn to have confidence in himself, having learnt, as it were, the arts and skills of navigation so that he may be able to voyage on alone and even become a voyager leading others into exciting and fertile fields of creative activity.

Perhaps one can liken the ideal University course to a series of explorations by helicopter over some vast new terrain where experienced guides can point out the interesting and unique features of the landscape and at intervals land a small party in some specially favourable area that entices to a closer inspection and evaluation. I can think of this as a model for what a good lecture should be, namely a helicopter journey introducing the main features of a field and finally the display in detail of some exciting and new features that the lecturer has been personally involved in.

It is important that this introduction of the young student to the third world be not done in a too hackneyed and rigid manner. Furthermore, we must not be intellectually arrogant. We must face students honestly, admitting the grave limitations of our knowledge, and at the same time revealing the exciting adventure of questing into the unknown. This adventure has its great successes, but each such achievement extends the vistas of the unknowns. It has been well said that a professor should appear before his students as an ignorant man learning.

I have always adopted the attitude in my lectures that the students can themselves read up much of the background material and also the rather uninteresting fields that it is necessary for them to know about, but which themselves do not merit the personal presentation in a lecture. The lecture should be kept for the more adventurous aspects of the project so that the students can sense the on-going academic struggles which are always occurring in lively fields of scholarship. America's greatest teacher of chemistry, GILBERT NEWTON LEWIS, deliberately restricted his course to the interesting fields of chemistry.

I feel that some of the student criticisms are justified in so far as they regard the practice of University teaching as being a merely unimaginative presentation of standard subject matter that often is badly done and should be relegated to supplementary reading courses. It is important to consider the University as being primarily concerned in giving the young student the opportunity to discover for himself, in the immensity of World 3, the features that are specially tuned in to his interests and imagination. If this is to be done at all satisfactorily, then the course arrangement should be flexible and the teachers should be themselves creative minds able to arouse a sense of adventure in the students.

The Student Discontents

My criticism of the present student unrest is that it is so wrongly directed. It has almost no intellectual content and eschews rationality, aiming only at emotional satisfactions in life and a sense of animal togetherness. Professor J. MCAULEY (1969) elaborates features of student activism that reveal it to be a left-wing Fascism in the making:

> "The revolt against liberalism and constitutional government; the cult of youthfulness; the rejection of the 'bureaucratic' and 'materialistic' institutions in favour of an unprogrammed future to be marked by spontaneity of the deed; the sense of belonging to a charismatic elite whose mandate to destroy the rotten fraud of bourgeois liberalism is by self-appointment, and whose style is total and abusive contempt for ordinary people; the low level of theoretization, which is in any case anti-rational in tendency; and also one might add the mania for dressing-up in costumes, for play-acting—and the fantastic self-importance assumed by psychically flawed unstable persons."

Almost no effort is made to evaluate courses on their academic content but only on their relationship to some immediate social problems. The most glaring of all of these tragedies is the efforts of the black students to have courses devoted to black history, culture, and power and not to mathematics, engineering, medical science, and the physical sciences. The late Tom Moyba of Kenya criticized much of the education provided in the past for the emerging peoples of African colonies as being wrongly directed to law and the social sciences, and not to the hard subjects of science and technology such as physics, chemistry, the medical sciences and engineering. Yet the same erroneous developments are being demanded by Black Power—courses that train for political activism and not the courses that would give black people the best opportunities in this scientific and technological age.

The "soft" social sciences provide a demonstration of the degree of deterioration of the ideals of scholarship and creativity that until recently have characterized all universities worthy of the name. As an example I quote from an article, "The Myth of the Free Scholar" by Dr. WOLFE (1969), an Assistant Professor of Political Science at the State University College of New York at Old Westbury. It is written in the style of smart cynicism that is affected by young academics with advanced pretensions and has been specially selected for reprinting in the University Review of the State University of New York. If only partly true, it is a severe indictment of political science as an academic discipline. It contrasts with my attempt to outline the stages in the training of a scientist in Chapter IX, as may be illustrated by this first quotation:

"I would argue that our universities are in disastrous condition because of the laissez-faire pluralism which has been allowed to exist there. My indictments are essentially two. The first is that the adherence to a spurious free-scholar model has obscured reality to the point where academic establishments can almost totally define what will pass for scholarly research. This adherence is used to justify the power these individuals have and serves to remove from the university people who do not cooperate. The result is that the university differs from other institutions only in being a little more hypocritical. Secondly, academic pluralism has insured the predominance of conservative scholarship in a conservative society."

As an alternative WOLFE states:

"I would advance the thesis that as socialism is the major alternative to laissez-faire, the social university is the alternative to what we have now. The social university is not primarily concerned with the abstract pursuit of scholarship, but with the utilization of knowledge obtained through scholarship to obtain social change. Therefore, it does not recognize the right of its members to do anything they wish under the name of academic freedom; instead it assumes that all its members are committed to social change. To give an example, a course in riot control would simply be declared out of place in such a university, while a course in methods of rioting might be perfectly appropriate."

It is evident enough that this "social university" is nothing more than a political powerhouse run on totalitarian lines. These extremes of radicalism and intolerance by junior faculty in the "soft" social sciences demonstrate the degree of disorder in universities that has been so effectively expressed by MCAULEY (1969).

"The totalitarian drift of the new radicalism is revealed in its will to politicize everything, to convert everything into a vehicle for revolutionary action: specifically, to convert the university from a seat of scholarship into a privileged sanctuary for the prosecution of guerilla warfare against the community and the government, a launching-pad for political missiles such as demonstrations, riots, subscriptions for aid to the enemy, and the organization of 'civil disobedience.' The essential immorality of this must be clearly stated. Every institution implies a moral contract, and to violate that contract is to commit an act of injustice against one's fellows."

Yet so many middle-aged parents refer to the disordered and rebellious youth as just "kids," as if they were suffering from a delayed adolescence. This seems to result from the parents being conditioned by their children to believe in fairy tales about them. Whenever I hear this talking of the "kids," I writhe because the emotionally slanted stories just reek with sentimentality! There is an illusion that the new generation has high moral purpose and dedication—some have—but far too many display arrogance and pride in their assertive claim to be superior to the older generation. They do not realize that arrogance and pride are the ultimate evils. Moreover the claim to do their "thing" is a claim to selfish gratification, that so often finds its satisfaction in drug addiction with its sequels of personal disintegration and degeneration. Worse still there is the terrible evil of tempting the innocent and inexperienced to go on psychedelic trips that can have disastrous and even fatal consequences. Modern "seers" preach to the students sermons that they want to hear about peace and the evil of war and killing, but I have yet to hear these "seers" preach against the evil of drug addiction and in particular against the evil of tempting the innocent.

The student discontents seem to me to be analogous to mountain sickness or some other deficiency disease. The students sense that something is amiss in the education that should fit them for their journey into the future, but they have no understanding of the nature of the disease that is responsible for these symptoms that they sense. The epidemic attack of an unknown disease leads to the confident diagnosis and treatment by corrupt and misguided seekers after power. What we need is not this counterfeit therapy by witch doctors and academic quacks, but wise, patient and inspired efforts to understand the disorder and so to treat it with skill and success.

Education for the Future

Universities are now too large and too disturbed by the masses. The Homo Socialis as defined by CHARLES FONTENAY (1968) has taken over. For example we have the horrid spectacle of some hippie student stating in a student discussion with faculty at Buffalo: "This is as much my university as MEYERSON's and REGAN's"—(naming the President and the Acting President). The original function of the universities as sanctuaries for scholarship and research has been submerged by the requirement that they become service institutions for society, providing a technical or professional training for the great mass of the youth and also being centers for the diagnosis and treatment of the disorders afflicting society today—such as the growth of violence, the decay of cities, the pollution of air, land and water, the dehumanization of existence in a mechanized society. The development of adaptive arrangements might have provided a means whereby this great burden could have been tolerated without destruction of the vital role of universities, namely their function as academic centers devoted to assimilating and transmitting the cultural heritage of mankind and as centers for the creative activities of the intellectual elite in the humanities and the sciences. But it now seems to be too late to hope for such adaptation, which would require much more time for peaceful evolution than is now available. One can think of the development of institutions associated with Universities much as are hospitals with medical Schools.

Throughout almost all of the free world there is a conspiracy to destroy universities by violent disruptions staging the presentation of non-negotiable demands. Great universities have been virtually paralysed by disorder, and those not yet attacked are so fearful that academic work suffers greatly. Such adaptations as are attempted are regarded by the revolutionary students and faculty as being merely appeasement by the already defeated universities; hence there will be further non-negotiable demands until universities may come to resemble the great cultural institutions of the past in name only. It seems certain that the most tragic loss has been the academic freedom that has been the outstanding treasure of universities. Violence has replaced the rational discussion that hitherto has so uniquely distinguished universities.

What then is to be done? I number myself amongst those academics who have no taste for this battle, and who see that it would paralyse their creative work to no good purpose. We see the so-called universities inundated by the destructive tide of student dissent. Yet we do see clearer than ever before the necessity for islands of academic peace and freedom, where scholarship and creative work can continue, else our civilization is lost in the progressive and universal degradation that occurs if man in the mass dominates all human activity. This determining factor in the fall of civilizations has been well displayed by CHARLES FONTENAY (1968) in his book "Epistle to the Babylonians." A civilization decays to oblivion when *Homo Individualis*, creative man, no longer is accepted for the leadership of the mass of men, *Homo Socialis*. So it has been so many times before, and one does not need to be a pessimist to fear that on the stage of history our epoch is already well past the high noon of brilliant illumination. Have not our so-called artists made this all too painfully clear by the miserable traumatic experiences they provide for our vision and our hearing?

I do not counsel despair, but I want to stress the desperate urgency of the present historical situation and the necessity for new creative efforts by our society in order that Homo Individualis may have the chance to restore to health our diseased and declining civilization. I believe that the great mega-universities are so stressed and threatened that many no longer can be considered appropriate sanctuaries for groups of Homo Individualis that require relative peace for the study and contemplation that leads to great creative achievements. And a new upsurge of creative activity could again give man faith in himself so that he can be successful in the great task of building a new civilization as we move from the twentieth to the twenty-first century.

There are already small institutions for the intellectual elite that are known as "Institutes for Advanced Study." There is a famous one at Princeton, and Rockefeller University is another. More of these are needed for the hard-pressed intellectuals. I worked for some fourteen years under attractive conditions at the Institute for Advanced Studies of the Australian National University at Canberra. In retrospect, and as demonstrated by our research performance, I can have confidence in asserting that we were provided with ideal conditions. There is urgent need for more of these so-called "ivory towers," else all civilization is lost in the "market place," which is the symbolic location of such a large part of a modern university. These institutes need not be isolated, but could be integral parts of existing Universities or juxtaposed as at Princeton. It can be hoped that from such academic oases there will flow to the great universities regenerative ideas for the future that it is our task to create.

According to the founder of the influential "Prospective" school of Philosophy, GASTON BERGER (1967a, 1967b), we should not regard the future as being an extrapolation of present trends but rather as a problem to be faced. The problem of education is to prepare man for what has not happened yet. But of course man, himself, is largely instrumental in bringing about this, as yet unknown, future. The great merit of this statement is that it presupposes a flexibility in approach. It further presupposes that education is not concentrated in the soft social sciences. On the contrary the aim of "prospective" education must be to train experienced thinkers to venture into the future from the secure base provided by an understanding and appreciation of the history of mankind in the long and heroic struggle upwards from barbarism. Sensitivity and wisdom would guide creativity to give the utmost of flexibility with the minimum of dogmatism. Only the basic academic disciplines can give the essential framework on which the future should be built.

This concept brings us back to the theme of this chapter—namely education and the world of objective knowledge. Each age has to know and evaluate the inheritance of World 3 deriving from past ages. The essential role of universities and their associated Institutes of Advanced Studies is to give the students of each generation the opportunity to sense, to experience and to immerse themselves in the resources and riches of World 3 that are most attractive and desirable to them. It will be appreciated that this is no soft and easy option. It involves great dedication and seriousness of purpose to become so involved, but that is the price of the on-going adventure that is ours when we live a life tuned in to the great vistas of beauty, truth and goodness that are the heritage of civilized man.

Chapter XII

Epilogue

It can be claimed that the philosophical position outlined in this book has the merit of encompassing in principle all experience. Also it has the merit of being based on the present scientific understanding of the brain. Admittedly, the philosophical adventures are at a very elementary level, but I believe that they are consistent within themselves and that the metaphysical suppositions are adequate for the conceptual developments. Such features have been conspicuously lacking in all of the materialist and behaviourist philosophies, which arbitrarily reject much of experience, and which are based on initial metaphysical assumptions, though later metaphysics is repudiated.

We know that persons are not just behaving units, because we can look within ourselves and see our own conscious individuality. So I warn you against the philosophies that exclusively claim to build upon the nature of man as a behaving being and that lead to some caricature of man, to some computer or cybernetic or robot man. To many, such philosophies provide satisfactory explanations of man as viewed from the outside, but they fail abysmally when applied to man as seen from the inside, which is the privileged position each of us has in respect of his own self.

In his recent book, "So Human an Animal," RENÉ DUBOS (1968) has stated the problem of man in much the same way as in Chapter I.

> "The most poignant problem of modern life is probably man's feeling that life has lost significance. The ancient religious and social creeds are being eroded by scientific knowledge and by the absurdity of world events. As a result, the expression 'God is dead' is widely used in both theological and secular circles. Since the concept of God symbolized the totality of creation, man now remains without anchor. Those who affirm the death of God imply thereby the death of traditional man—whose life derived significance from his relation to the rest of the cosmos. The search for significance, the formulation of new meanings for the words God and Man, may be the most worthwhile pursuit in the age of anxiety and alienation."

This word "alienation" has been much used in recent times to describe man's predicament as he recognizes that he no longer has the faith to live his life with hope. There were outlined in Chapter I several causes for the present sickness of mankind that expresses itself in this anxiety and alienation.

The title of this book, "Facing Reality," refers to personal reality and the book is concerned with the attempt of each person to face up to his own personal existence as a unique conscious self—as is particularly treated in Chapters IV, V, VI and X.

It emerges in Chapters II and III that there are tremendous unknowns in the scientific effort to understand the structure and mode of operation of the brain and so to correlate this with the conscious experiences that in some manner are dependent on activities in the fantastically complex neuronal organization of the brain (Chapters IV, V, VI and VIII). Since science is itself dependent on brain action, one may well wonder if there will ever be a complete scientific description of the brain. As has often been pointed out, there seems to be a paradox involved in a brain attempting to understand itself. Nevertheless, such is the power of modern scientific techniques, that we can anticipate great discoveries in the field of the brain sciences. There will be a great increase in our understanding, for example in respect of memory and the control of movement, but such fundamental problems as mind-brain liaison in perception and free will will be beyond any conceivable investigation. It seems that these problems can only be solved at the expense of a complete transformation of science in an as yet unimaginable manner (Chapters IV, V and VIII).

Doubtless, by many, the reality to be faced is the reality of the confrontation of world powers armed with nuclear weapons that if unleashed would give terrible destruction. And one of these powers is a totalitarian tyranny and the only aggressive imperialism in the world today. But one must not cringe before this terrible reality. Facing that reality has been the theme of many books. Yet the reality that is the theme of this book is more stark, for it relates to the conscious existence that inevitably ends in death for tyrants as well as for the enslaved and the free. It is this reality that each of us has to face—or refuse to face!

Towards the end of his great book, "Man on his Nature," SHERRINGTON (1940) summed up the journey that his audience had experienced during the Gifford Lectures at the University of Edinburgh during 1937 and 1938.

"We traced how, it would seem, we are so fashioned that our world, which is our experience and is one world, is a diune world, a world of outlook and of inlook, of the experienced perceptible and of the experienced imperceptible. This world with all its sweep of content and extent taxes utterance to indicate. Yet it is given us in so far to seize it, and as one coherent harmony. More; it is revealing to us the 'values', as Truth, Charity, Beauty. Surely these are compensation to us for much. And will not this compensation grow? Charity will grow; Truth grows; and even as Truth so Beauty. Music as her ear grows finer embraces what once were discords. The mind which began by being one thing has truly

—as so often in evolution—gone on to being another thing. Even should mind in the cataclysm of Nature be doomed to disappear and man's mind with it, man will have had his compensation: to have glimpsed a coherent world and himself as item in it. To have heard for a moment a harmony wherein he is a note. And to listen to a harmony is to commune with its Composer?"

We can have hope as we recognize and appreciate the wonder and mystery of our existence as experiencing selves. Mankind would be cured of his alienation if that message could be expressed with all the authority of scientists and philosophers as well as with the imaginative insights of artists. In this book I express my efforts to understand a human person, namely myself, as an experiencing being. I offer it in the hope that it may help man to discover a way out of his alienation and to face up to the terrible and wonderful reality of his existence—with courage and faith and hope. I pray that man may develop a transforming faith in the meaning and significance of this wonderful, even unbelievable, adventure that each of us is given on this lovely and salubrious earth of ours, that is itself a mere grain in the infinite cosmos of galaxies. Because of the mystery of our being as unique self-conscious existences, we can have hope as we set our own soft, sensitive and fleeting personal experience against the terror and immensity of illimitable space and time. Are we not participants in the meaning, where there is else no meaning? Do we not experience and delight in fellowship, joy, harmony, truth, love and beauty, where there else is only the mindless universe?

References

[Numbers in square brackets at end of each entry indicate the pages on which it is cited.]

ABELSON, P.H.: Paleobiochemistry, Scientific American, **195**, 83–92 (1956). [88]

ADLER, M.J.: The difference of man and the difference it makes. New York-Chicago-San Francisco: Holt, Reinhart and Winston 1967. [152, 170]

ADRIAN, E.D.: The physical background of perception, 95 pp. Oxford: The Clarendon Press 1947. [91]

AGRANOFF, B.W.: Agents that block memory. In: The neurosciences, ed. by G.C. Quarton, T. Melnechuk and F.O. Schmitt. New York: Rockefeller University Press 1967. [40, 41]

ALBE-FESSARD, D., FESSARD, A.: Thalamic integrations and their consequences at the telencephalic level. In: Progress in brain research, vol. 1, Brain mechanisms, ed. by G. Moruzzi, A. Fessard and H.H. Jasper, p. 115–154. Amsterdam: Elsevier Publ. Co. 1963. [24]

ANDERSEN, P., HOLMQVIST, B., VOORHOEVE, P.E.: Entorhinal activation of dentate granule cells. Acta physiol. scand. **66**, 448–460 (1966). [31]

BAILLIE, J.: And the life everlasting. London: Oxford University Press 1934. [174]

BARONDES, S.H.: The mirror focus and long-term memory storage. In: Basic mechanisms of the epilepsies, ed. by H.H. Jasper, A.A. Ward and A. Pope. Boston: Little, Brown & Company 1969. [40]

— COHEN, H.D.: Delayed and sustained effect of acetoxycycloheximide on memory in mice. Proc. nat. Acad. Sci. (Wash.) **58**, 157 (1967). [40]

— — Memory impairment after subcutaneous injection of acetoxycycloheximide. Science **160**, 556 (1968). [40]

BASMAJIAN, J.V.: Control and training of individual motor units. Science **141**, 440–441 (1963). [119]

BELOFF, J.: The existence of mind. London: MacGibbon & Kee 1962. [53, 64, 152]

BERGER, G.: Sciences humaines et prévision, p. 16–26. Étapes de la Prospective. Paris: Presses Universitaires de France 1967a. [186]

— L'Attitude prospective, p. 27–34. Étapes de la Prospective. Paris: Presses Universitaires de France 1967b. [186]

BLACKETT, P.M.S.: Address of the President. Proc. roy. Soc. B **196** V–XVIII (1968). [148, 149]

BLISS, T.V.P., LØMO, T.: Plasticity in a monosynaptic cortical pathway. J. Physiol. (Lond.) **207**, 21 P (1970) and unpublished observations. [31, 32]

BRAIN, W.R.: Mind, perception and science, 90 pp. Oxford: Blackwell Scientific Publications 1951. [69]

BREMER, F.: Neurophysiological correlates of mental unity. In: Brain and conscious experience, p. 283–297, ed. by J.C. Eccles. Berlin-Heidelberg-New York: Springer 1966. [73]

BROOKS, C., McC., ECCLES, J.C.: An electrical hypothesis of central inhibition. Nature (Lond.) **159**, 760–764 (1947a). [128]

BROWN, J.: Short-term memory. Brit. med. Bull. **20**, 8–11 (1964). [25]

BURNS, B.D.: Some properties of the isolated cerebral cortex of the unanaesthetized cat. J. Physiol. (Lond.) **112**, 156–175 (1951). [17, 120]

— The mammalian cerebral cortex, 119 pp. London: Edward Arnold Ltd. 1958. [17, 25]

BUSER, P., IMBERT, M.: Sensory projections to the motor cortex in cats: a microelectrode study. In: Sensory communication. Symposium on principles of sensory communication, ed. by W.A. Rosenblith, p. 607–626. London: John Wiley & Sons, Inc. 1961. [24]

CALVIN, M.: Chemical evolution. Eugene, Oregon: University of Oregon Press 1961. [88]

— Chemical evolution. London: Oxford University Press 1969. [88]

CASSIRER, E.: An essay on man. New Haven: Yale University Press 1944. [170]

CHAMBERLAIN, T.J., HALICK, P., GERARD, R.W.: Fixation of experience in the rat spinal cord. J. Neurophysiol. **26**, 662–673 (1963). [39]

CHORON, J.: Death and western thought. London: Ballica-Macmillan 1963. [174]

— Modern man and mortality. New York: The Macmillan Company 1964. [174]

COLONNIER, M.L.: The structural design of the neocortex. In: Brain and conscious experience, p. 1–23, ed. by J.C. Eccles. Berlin-Heidelberg-New York: Springer 1966. [16, 37]

— Synaptic patterns on different cell types in the different laminae of the cat visual cortex. An electron microscope study. Brain Res. **9**, 268–287 (1968). [15, 33]

— ROSSIGNOL, S.: Heterogeneity of the cerebral cortex. In: Basic mechanisms of the epilepsies, ed. by H.H. Jasper, A.A. Ward and A. Pope. Boston: Little, Brown & Company 1969. [15]

COOMBS, J.S., CURTIS, D.R., ECCLES, J.C.: The generation of impulses in motoneurones. J. Physiol. (Lond.) **139**, 232–249 (1957). [14]

— ECCLES, J.C., FATT, P.: The inhibitory suppression of reflex discharges from motoneurones. J. Physiol. (Lond.) **130**, 396–413 (1955). [14]

COWAN, J.D.: Redundant automata as models of neuron assemblies. In: Information processing in the nervous system, p. 397–403, ed. by R.W. Gerard and J.W. Duyff. Amsterdam: Excerpta Medica 1964. [17]

CRAWFORD, B.H.: Visual adaptation in relation to brief conditioning stimuli. Proc. roy. Soc. B **134**, 283–302 (1947). [56, 71, 160]

CREUTZFELDT, O., FUSTER, J.M., HERZ, A., STRASCHILL, M.: Some problems of information transmission in the visual system. In: Brain and conscious experience, ed. by J.C. Eccles, p. 138–160. Berlin-Heidelberg-New York: Springer 1966. [22]

CRICK, F.: Of molecules and man, 117 p. Seattle: University of Washington Press 1966. [7, 173]

CURTIS, D.R., ECCLES, J.C.: The time courses of excitatory and inhibitory synaptic actions. J. Physiol. (Lond.) **145**, 529–546 (1959). [14]

— — Synaptic action during and after repetitive stimulation. J. Physiol. (Lond.) **150**, 374–398 (1960). [31]

DALE, H.H.: The beginnings and the prospects neurohumoral transmission. Pharm. Rev. **6**, 7–13 (1954). [106]

DESCARTES, R.: Philosophical works, trans. E.S. Haldane and G.R.T. Ross. Cambridge: Cambridge University Press 1931. [55, 171]

DEUTSCH, M.: Evidence and inference in nuclear research. In: Evidence and inference, ed. by D. Lerner. Glencoe, Ill.: The Free Press 1959. [104]

DEWEY, J.: Psychology. (Third ed.) New York: American 1898. [49]

DINGMAN, W., SPORN, N.B.: Molecular theories of memory. Science **144**, 26–29 (1964). [39]

DIXON, W. M.: The human situation: problems of life and destiny. London: Arnold & Co. 1937. [174]

DOBZHANSKY, T.: Mankind evolving: The evolution of the human species. New Haven: Yale University Press 1962. [81, 85, 89, 92]

— The biology of ultimate concern. New York: New American Library 1967. [62, 81, 85, 87, 90–92, 94, 97, 98, 101, 152, 172]

— The pattern of human evolution. In: The uniqueness of man, ed. by J. D. Roslansky. Amsterdam: North-Holland Publishing Company 1969. [46, 62, 172]

DROZ, G., BARONDES, H.: Nerve Endings: Rapid appearance of labeled protein shown by electron microscope radioautography. Science **165**, 1131–1133 (1969). [41]

DUBNER, R., RUTLEDGE, L. T.: Recording and analysis of converging input upon neurons in cat association cortex. J. Neurophysiol. **27**, 620–634 (1964). [22]

DUBOS, R.: So human an animal. New York: Charles Scribner's Sons 1968. [1, 188]

ECCLES, J. C.: Man and freedom. In: Twentieth Century, Melbourne, Australia **5**, 23 (1947). [130]

— The neurophysiological basis of mind: The principles of neurophysiology, p. 314. Oxford: Clarendon Press 1953. [28, 58, 121, 124, 128, 163]

— The physiology of imagination. Scientific American **199**, 135–146 (1958). [21, 23]

— The effects of use and disuse on synaptic function. In: Brain mechanisms and learning, ed. by J. F. Delafresnaye, p. 335–352. Oxford: Blackwell Scientific Publications, 1961. [28]

— The physiology of synapses, 316 p. Berlin-Göttingen-Heidelberg: Springer 1964. [13, 28, 106, 110–112, 148, 149]

— The brain and the unity of conscious experience. (Eddington Lecture.) London: Cambridge University Press 1965a. [1, 62, 63, 174, 175]

— The brain and the person. Melbourne: Australian Broadcasting Commission 1965b. [1]

— Some observations on the strategy of neurophysiological research. In: Nerve as a tissue, ed. by Kaare Rodahl, p. 445–455. New York: Harper & Row 1966a. [117]

— Conscious experience and memory. In: Brain and conscious experience, p. 314–344, ed. by. J. C. Eccles. Berlin-Heidelberg-New York: Springer 1966b. [21, 25, 28, 39]

— Conscious experience and memory. Rec. Advanc. biol. Psychiat. **8**, 235–256 (1966c). [25]

— Ionic mechanisms of excitatory and inhibitory synaptic action. Ann. N.Y. Acad. Sci. **137**, 473–494 (1966d). [13, 111]

— Brain and the development of the human person. Impact (UNESCO, Paris) **16**, 93–112 (1966e). [180]

— Evolution and the conscious self. In: The human mind, ed. by J. D. Roslansky. Amsterdam: North-Holland Publishing 1967. [62, 85, 174].

— The importance of brain research for the educational, cultural and scientific future of mankind. Perspect. Biol. Med. **12**, 61–68 (1968). [1]

— The experiencing self. In: The uniqueness of man, ed. by J. D. Roslansky. Amsterdam: North-Holland Publishing Co. 1969a. [1, 44, 174]

— The inhibitory pathways of the central nervous system. Liverpool, England: Liverpool University Press 1969b. [15, 108, 113]

— The necessity of freedom for the free-flowering of science. Dunning Trust Lecture, Queens University. Queens Quarterly (1969c). [135]

— The dynamic loop hypothesis of movement control. In: Information processing in the nervous system, ed. K. N. Leibovic. Berlin-Heidelberg-New York: Springer 1970a. [50, 107]

— Neurogenesis and morphogenesis in the cerebellar cortex. Proc. nat. Acad. Sci. (Wash.) **66**, 295–301 (1970b). [113]

ECCLES, J.C., HUBBARD, J.I., OSCARSSON, O.: Intracellular recording from cells of the ventral spinocerebellar tract. J. Physiol. (Lond.) **158**, 486–516 (1961). [29]

— ITO, M., SZENTÁGOTHAI, J.: The cerebellum as a neuronal machine. Berlin-Heidelberg-New York: Springer 1967. [37]

— MCINTYRE, A.K.: The effects of disuse of activity on mammalian spinal reflexes. J. Physiol. (Lond.) **121**, 492–516 (1953). [31]

EDDINGTON, A.S.: Science and the unseen world. London: George Allen & Unwin Ltd. 1929. [63]

— New pathways in science, 230 p. Cambridge University Press 1935. [125]

— The philosophy of physical science. London: Cambridge University Press 1939. [63, 65, 125, 126, 152].

EIGEN, M.: Chemical means of information storage, and readout in biological systems. Neurosci. Res. Progr. Bull. Cambridge, Mass.: M.I.T. Press 11–22 (1964). [25]

EVARTS, E.V.: Effects of sleep and waking on activity of single units in the unrestrained cat. In: The nature of sleep, ed. by G.E.W. Wolstenholme and M. O'Connor. London: J. & A. Churchill Ltd. 1961. [22, 161]

— Activity of neurons in visual cortex of the cat during sleep with low voltage fast EEG activity. J. Neurophysiol. **25**, 812–816 (1962). [22, 161]

— Temporal patterns of discharge of pyramidal tract neurons during sleep and waking in the monkey. J. Neurophysiol. **27**, 152–171 (1964). [22, 72, 161]

FEIGL, H.: The "mental" and the "physical". Minneapolis, Minnesota: University of Minnesota Press 1967. [159, 173]

FESSARD, A.: Mechanisms of nervous integration and conscious experience. In: Brain mechanisms and consciousness, p. 200–236, ed. by J.F. Delafresnaye. Oxford: Blackwell 1954. [17]

— The role of neuronal networks in sensory communications within the brain. In: Sensory communication. Symposium on principles of sensory communication, ed. W.A. Rosenblith, p. 585–606. London: John Wiley & Sons, Inc. 1961. [17, 21, 22, 54, 58]

FONTENAY, C.L.: Epistle to the Babylonians. An essay on the natural inequality of man. Knoxville: University of Tennessee Press 1968. [180, 184, 185]

FOX, S.W.: Simulated natural experiments in spontaneous organization of morphological units from proteinoid. The origins of prebiological systems and of their molecular matrices (editor S.W. Fox), p. 361–373. 1964: Academic Press, New York. [88]

FROMM, E.: The heart of man, its genius for good and evil. New York: Harper 1964. [90]

FURNESS, W.H.: Observations on the mentality of chimpanzees and orangutans. Proc. Amer. Phil. Soc. **55**, 281–290 (1916). [95]

GASTAUT, H.: Some aspects of the neurophysiological basis of conditioned reflexes and behavior. In: Neurological basis of behavior. London: J. & A. Churchill Ltd. 1958 [42]

GERARD, R.W.: Physiology and psychiatry. Amer. J. Psychiat. **106**, 161–173 (1949). [25]

GLASSMAN, E.: Some considerations of the effects of short term learning on the incorporation of uridine into RNA and polysomes of mouse brain. In: The future of the brain sciences, ed. by S. Bogoch. New York: Plenum Press 1969. [41]

GOMULICKI, B.R.: The development and present status of the trace theory of memory. Cambridge: Cambridge University Press 1953. [26]

GRANIT, R.: Receptors and sensory perception. New Haven: Yale University Press 1955. [161]

— Sensory mechanisms in perception. In: Brain and Conscious Experience. Ed. by J.C. Eccles. Heidelberg: Springer-Verlag 1966. [22]

— The development of retinal neurophysiology, pp. 232–241. In: Les Prix Nobel en 1967. Stockholm: Nobel Foundation 1968. [161]

HAMLYN, L.H.: An electron microscope study of pyramidal neurons in the Ammon's horn of the rabbit. J. Anat. (Lond.) **97**, 189–201 (1963). [11–13]
HÁMORI, J., SZENTÁGOTHAI, J.: The "crossing over" synapse. An electron microscope study of the molecular layer in the cerebellar cortex. Acta biol. Acad. Sci hung. **15**, 95–117 (1964). [38]
HARDY, A.: The living stream. Evolution and man. New York: Harper & Row 1965. [152, 166]
HARRIS, C.S.: Perceptual adaptation to inverted, reversed and displaced vision. Psychol. Rev. **72**, 419–444 (1965). [50]
HAYES, H.J., HAYES, C.: The cultural capacity of chimpanzee. Hum. Biol. **26**, 288–303 (1954). [95, 96]
HEBB, D.O.: The organization of behaviour. New York: John Wiley & Sons 1949. [25, 28, 41]
HEBERER, G.: The descent of man and the present fossil record. Cold Spr. Harb. Symp. quant. Biol. **24**, 235–244 (1959). [92]
HELD, R., HEIN, A.: Movement-produced stimulation in the development of visually guided behavior. J. comp. Physiol. Psychol. **56**, 872–876 (1963). [66, 67]
HINSHELWOOD, C.N.: The vision of nature. 15th Eddington Memorial Lecture. London: Cambridge University Press 1962. [47, 152]
HOERNER, S. von: The general limits of space travel. In: Interstellar communication, ed. by A.G.W. Cameron, p. 144–159. New York and Amsterdam: W.A. Benjamin, Inc. 1963. [99]
HOLTON, G.: Science and new styles of thought. The Graduate Journal. The University of Texas, **7**, 399–422 (1967). [104]
HUBEL, D.H.: The visual cortex of the brain. New York: Scientific American 1963. [154, 158]
— WIESEL, T.N.: Receptive fields, binocular interaction and functional architecture in the cat's visual cortex. J. Physiol. (Lond.) **160**, 106–154 (1962). [22, 156]
— — Shape and arrangement of columns in the cat's striate cortex. J. Physiol. (Lond.) **165**, 559–568 (1963). [22, 156]
— — Receptive fields and functional architecture in two non-striate visual areas (18 and 19) of the cat. J. Neurophysiol. **28**, 229–289 (1965). [22, 157, 160]
HÜGEL, F. von: Eternal Life. A study of its implications and applications. Edinburgh: T. & T. Clark 1912. [174]
HUXLEY, J.: Higher and lower organization in evolution. J. Roy. Coll. Surg. Edinb. **7**, 163–179 (1962). [51, 91]
HYDÉN, H.: Biochemical changes in glial cells and nerve cells at varying activity. Proc. Fourth Inter. Congr. Biochemistry, vol. 3, p. 64–89, ed. by O. Hoffman-Osternhof. London: Pergamon Press 1959. [26, 41]
— Introductory remarks to the session on memory processes. Neurosciences Research Program Bull. Cambridge, Mass.: M.I.T. Press 23–38 (1964). [26, 27, 39, 41]
— Activation of nuclear RNA in neurons and glia in learning. In: Anatomy of memory, ed. by D.P. Kimble. Palo Alto, California: Science and Behavior Books, Inc. 1965. [26]
— Biochemical changes accompanying learning. In: The neurosciences, ed. by G.C. Quarton, T. Melnechuk and F.O. Schmitt. New York: Rockefeller University Press 1967. [26, 27, 41]
JAKI, S.L.: Brain, mind and computers. New York: Herder & Herder 1969. [152, 171]
JASPER, H.H.: Pathophysiological studies of brain mechanisms in different states of consciousness. In: Brain and conscious experience, p. 256–282, ed. by J.C. Eccles. Berlin-Heidelberg-New York: Springer 1966. [71, 160]

JASPER, H.H., RICCI, G.F., DOANE, B.: Patterns of cortical neuronal discharge during conditioned responses in monkeys. In: Neurological basis of behavior. London: J. & A. Churchill, Ltd. 1958. [42]

JENNINGS, H.S.: The biological basis of human nature. New York: W.W. Norton & Company, Inc. 1930. [80–83]

JUNG, R.: Neuronal integration in the visual cortex and its significance for visual information. In: Sensory communication. Symposium on principles of sensory communication, ed. by W.A. Rosenblith, p. 627–674. New York: John Wiley & Sons, Inc. 1961. [22]

— KORNHUBER, H.H., FONSECA, J.S. da: Multisensory convergence on cortical neurons. Neuronal effects of visual, acoustic and vestibular stimuli in the superior convolutions of the cat's cortex. In: Progress in brain research, vol., 1. Brain mechanisms, ed. by G. Moruzzi, A. Fessard and H.H. Jasper, p. 207–240. Amsterdam: Elsevier Publ. Co. 1963. [22]

KANDEL, E.R., SPENCER, W.A.: Cellular neurophysiological approaches in the study of learning. Physiol. Rev. **48**, 65–134 (1968). [26, 28, 31]

KELLOGG, W.N., KELLOGG, L.A.: The ape and the child. New York: McGraw Hill Book Company 1933. [95]

KERKUT, G.A., THOMAS, R.C.: The effect of anion injection and changes in the external potassium and chloride concentration on the reversal potentials of the IPSP and acetylcholine. Comp. Biochem. Physiol. **11**, 199–213 (1964). [110]

KLEITMAN, N.: The nature of dreaming. In: The nature of sleep, ed. by G.E.W. Wolstenholme and M. O'Connor. London: J. & A. Churchill, Ltd. 1961. [57, 161]

— Sleep and wakefulness, x + 552 p. Chicago: University of Chicago Press 1963. [57, 161]

KNEALE, W.: On having a mind. London: Cambridge University Press 1962. [64]

KOHLER, I.: Über Aufbau und Wandlungen der Wahrnehmungswelt. S.-B. öst. Akad. Wiss., philohist. Kl. **227**, 1–118 (1951). [49]

KORNHUBER, H.H., ASCHOFF, J.C.: Somatisch-vestibuläre Integration an Neuronen des motorischen Cortex. Naturwissenschaften **51**, 62–63 (1964). [22]

KROEBER, A.L.: The superorganic. Amer. Anthropol. **19**, 163–213 (1952). [96]

LACK, D.: Evolutionary theory and Christian belief. London: Methuen & Co., Ltd. 1961. [85, 91]

LANDGREN, S., PHILLIPS, C.G., PORTER, R.: Minimal synaptic actions of pyramidal impulses on some alpha motoneurones of the baboon's hand and forearm. J. Physiol. (Lond.) **161**, 91–111 (1962). [29, 30]

LANGER, S.K.: Philosophy in a new key. Cambridge, Mass.: Harvard University Press 1951. [95]

LASHLEY, K.S.: In search of the engram. Symp. Soc. exp. Biol. **4**, 454–482 (1950). [25, 28, 43, 76]

LIBET, B.: Brain stimulation and the threshold of conscious experience. In: Brain and conscious experience, p. 165–181, ed. by J.C. Eccles. Berlin-Heidelberg-New York: Springer 1966. [56, 69, 70, 72, 160]

LLOYD, D.P.C.: Post-tetanic potentiation of response in monosynaptic reflex pathways of the spinal cord. J. gen. Physiol. **33**, 147–170 (1949). [31]

LØMO, T.: Some properties of a cortical excitatory synapse. In: Excitatory synaptic mechanisms, ed. by P. Anderson and J. Jansen, Jr. Oslo: Oslo University Press 1970. [30, 31]

LORENTE DE NÓ, R.: Studies on the structure of the cerebral cortex. I. Area entorhinalis. J. Psychol. Neurol. (Lpz.) **45**, 381–438 (1933). [123]

— Studies on the structure of the cerebral cortex. II. Continuation of the study of the ammonic system. J. Psychol. Neurol. (Lpz.) **46**, 113–177 (1934). [123]

LORENTE DE NÓ, R.: Cerebral cortex: Architecture, intracortical connections, motor projections. Physiology of the nervous system, by J.F. Fulton, 2nd Edition, 614 pp. London: Oxford University Press 1943. [123]
MACKAY, D.M.: Cerebral organization and the conscious control of action. In: Brain and conscious experience, p. 422–445, ed. by J.C. Eccles. Berlin-Heidelberg-New York: Springer 1966. [4, 120]
MARGARIA, R.: The possibility of extraterrestial life. In XXII International Congress of Physiological Sciences. Suppl. to Proceedings, vol. 1, p. 108–113 (1962). [99]
MARR, D.: A theory of cerebellar cortex. J. Physiol. (Lond.) **202**, 437–470 (1969). [37]
MCAULEY, J.: Admirable Jeunesse? Quadrant (Sydney) **14**, 47–51 (1969). [182–184]
MILLER, S.L.: Production of some organic compounds under possible primitive Earth conditions. J. Amer. chem. Soc. **77**, 2351 (1955). [88]
— The mechanism of synthesis of amino acids by electric discharge. Biochim. biophys. Acta (Amst.) **23**, 488 (1957). [88]
MORRELL, F.: Lasting changes in synaptic organization produced by continuous neuronal bombardment. In: Brain mechanisms and learning, ed. by J.F. Delafresnaye. Oxford: Blackwell Scientific Publications 1961a. [39, 41]
— Electrophysiological contributions to the neural basis of learning. Physiol. Rev. **41**, 443–494 (1961b). [26]
— Physiology and histochemistry of the mirror focus. In: Basic mechanisms of the epilepsies, ed. by H.H. Jasper, A.A. Ward & A. Pope. Boston: Little, Brown & Co. 1969. [39, 41]
MORUZZI, G.: The functional significance of sleep with particular regard to the brain mechanisms underlying consciousness. In: Brain and conscious experience, p. 345–388, ed. by J.C. Eccles. Berlin-Heidelberg-New York: Springer 1966a. [56, 161]
— Brain plasticity. In: Brain and conscious experience, p. 555–560, ed. by J.C. Eccles. Berlin-Heidelberg-New York: Springer 1966b. [72, 73]
MOUNTCASTLE, V.B.: The neural replication of sensory events in the somatic afferent system. In: Brain and conscious experience, p. 85–115, ed. by J.C. Eccles. Berlin-Heidelberg-New York: Springer 1966a. [22, 72]
— The functional meaning of specific and nonspecific systems. In: Brain and conscious experience, p. 548–550, ed. by J.C. Eccles. Berlin-Heidelberg-New York: Springer 1966b. [72]
MYERS, R.E.: Corpus callosum and visual gnosis. In: Brain mechanisms and learning, ed. by J.F. Delafresnaye, p. 481–505. Oxford: Blackwell Scientific Publications 1961. [74]
ORWELL, G.: Nineteen eighty four. London: Harcourt, Bruce & Co. 1949. [149, 175]
PALAY, S.L.: The morphology of synapses in the central nervous system. Exp. Cell Res., Suppl. **5**, 275–293 (1958). [12]
PELIKAN, J.: The shape of death: life, death and immortality in the early Fathers. London: Macmillan & Co., Ltd. 1962. [174]
PENFIELD, W.: Speech and perception—the uncommitted cortex. In: Brain and conscious experience, p. 217–237, ed. by J.C. Eccles. Berlin-Heidelberg-New York: Springer 1966. [56, 74, 76]
— Engrams in the human brain. Proc. roy. Soc. Med. **61**, 831 (1968). [53]
— Epilepsy, neurophysiology, and some brain mechanisms related to consciousness. In: Basic mechanisms of the epilepsies, ed. by H.H. Jasper, A.A. Ward and A. Pope. Boston: Little, Brown & Company 1969. [74]
— JASPER, H.: Epilepsy and the functional anatomy of the human brain, p. 896. Boston: Little, Brown & Company 1954. [53, 69, 80]
— ROBERTS, L.: Speech and brain-mechanisms. Princeton, New Jersey: Princeton University Press 1959. [153]

PETTIGREW, J.D., NIKARA, T., BISHOP, P.O.: Responses to moving slits by single units in cat striate cortex. Exp. Brain Res. **6**, 373–390 (1968a). [156]

— — — Binocular interaction on single units in cat striate cortex: simultaneous stimulation by single moving slit with receptive fields in correspondence. Exp. Brain Res. **6**, 391–410 (1968b). [156]

PHILLIPS, C.G.: Changing concepts of the precentral motor area. In: Brain and conscious experience, p. 389–421, ed. by J.C. Eccles. Berlin-Heidelberg-New York: Springer 1966. [119]

POLANYI, M.: Personal knowledge. Towards a post-critical philosophy. London: Routledge & Kegan Paul 1958. [103, 138, 152]

— Science, faith and Society. Chicago: University of Chicago Press, Phoenix Books 1964. [138, 150]

— The tacit dimension. Garden City, New York: Doubleday & Company 1966. [59, 103, 138, 152, 173]

— Life transcending physics and chemistry. Chemical and Engineering News **45**, 54–66 (1967a). [59, 60, 173]

— Science and reality. Brit. J. Phil. Sci. **18**, 177–196 (1967b). [103]

— The growth of science in society. Minerva (Lond.) **5**, 533–545 (1967c). [145]

— Logic and psychology. Amer. Psychologist **23**, 27–43 (1968a). [103]

— Life's irreducible structure. Science **160**, 1308–1312 (1968b). [59, 60, 173]

POPPER, K.R.: Indeterminism in quantum physics and in classical physics. Brit. J. Phil. Sci. **1**, 117–133 (1950). [4, 120]

— The logic of scientific discovery, 480 p.p. London: Hutchinson 1959. [102, 142]

— Conjectures and refutations. The growth of scientific knowledge. New York and London: Basic Books 1962. [7, 115, 117, 142, 145, 146, 170]

— Science: problems, aims and responsibilities. Fed. Proc. **22**, 961–972 (1963). [103, 116]

— Epistemology without a knowing subject. In: Logic, methodology and philosophy of sciences. III, ed. by van Rootselaar and Staal. Amsterdam: North-Holland Publishing Company 1968a. [152, 164, 168, 176, 180, 181]

— On the theory of the objective mind. Akten des XIV. Internationalen Kongresses für Philosophie, vol. 1, Wien. (1968b). [152, 164, 169, 176]

PORTER, R.: Early facilitation at corticomotoneuronal synapses. J. Physiol. (Lond.) **207**, 733–745 (1970). [30]

POWELL, T.S.P., MOUNTCASTLE, V.B.: Some aspects of the functional organization of the cortex of the postcentral gyrus of the monkey: A correlation of findings obtained in a single unit analysis with cytoarchitecture. Bull. Johns Hopk. Hosp. **105**, 173–200 (1959). [22]

PURCELL, E.: Radioastronomy and communication through space. In: Interstellar communication, ed. by A.G.W. Cameron, p. 121–143. New York and Amsterdam: W.A. Benjamin, Inc. 1963. [99]

QUARTON, G.C.: The enhancement of learning by drugs and the transfer of learning by macromolecules. In: The neurosciences, ed. by G.C. Quarton, T. Melnechuk and F.O. Schmitt, p. 744–755. New York: Rockefeller University Press 1967. [39]

RAMÓN y CAJAL, S.: Histologie du Système Nerveux de L'Homme et des Vertébrés. II, 993 pp. Paris: Maloine. 1911. [9, 11, 13, 28]

RUIZ-MARCOS, A., VALVERDE, F.: The temporal evolution of the distribution of dendritic spines in the visual cortex of normal and dark raised mice. Exp. Brain Res. **8**, 284–294 (1969). [33]

RUSHTON, W.A.H.: Kinetics of cone pigments measured objectively on the living human fovea. Ann. N.Y. Acad. Sci. **74**, 291–304 (1958). [161]

RYLE, G.: The concept of mind, 334 pp. London: Hutchinson's University Library 1949. [64, 127]

SAWYER, D.B.: Personal Communication (1951). [122]
SCHMITT, F.O.: Molecular and ultrastructural correlates of function in neurons, neuronal nets, and the brain. Neurosci. Res. Progr. Bull. Cambridge, Mass.: M.I.T. Press 43–66 (1964). [26]
SCHRÖDINGER, E.: Science and humanism. London: Cambridge University Press 1951. [44, 63, 126]
— Mind and matter, p. 104. London: Cambridge University Press 1958. [52, 60, 61, 63, 100, 152]
SENDEN, M. VON: Space and sight. Translated by P. Heath. London: Methuen and Company, Ltd. 1960. [49]
SHERRINGTON, C.S.: The integrative action of the nervous system. New Haven and London: Yale University Press 1906. [15]
— Man on his nature, p. 413. London: Cambridge University Press 1940. [5, 54, 56, 65, 80, 91, 151, 159, 160, 161, 174, 189, 190]
— Foreword to 1947 Edition. The integrative action of the nervous system. Cambridge: Cambridge University Press 1947. [151, 155]
SHOLL, D.A.: The organization of the cerebral cortex. London: Methuen & Company, Ltd.; New York: John Wiley & Sons, Inc. 1956. [16, 154, 156, 159]
SIMPSON, G.G.: The principles of classification and a classification of mammals. Bull. Amer. Museum Nat. Hist. **85** (1945). [92]
— This view of life: the world of an evolutionist. New York: Harcourt, Brace & World 1964. [85, 89, 97]
SMITH, H.: Human versus artificial intelligence. In: The human mind, ed. by J.D. Roslansky. Amsterdam: North-Holland Publishing Co. 1967. [171]
SPERRY, R.W.: The great cerebral commissure. Scientific American **210**, 42–52 (1964). [74–76]
— Hemispheric interaction and the mind-brain problem. In: Brain and conscious experience, p. 298–313, ed. by J.C. Eccles. Berlin-Heidelberg-New York: Springer 1966. [74–79, 91]
STENT, G.S.: Induction and repression of enzyme synthesis, p. 152–161. In: The neurosciences, ed. by G.C. Quarton, T. Melnechuk and F.O. Schmitt. New York: Rockefeller University Press 1967. [39]
— The coming of the golden age. The American Museum of Natural History. Garden City, New York 1969. [7]
STRATTON, G.M.: Vision without inversion of retinal image. Psychol. Rev. **4**, 463–481 (1897). [49]
SZENTÁGOTHAI, J.: Structure-functional considerations of the cerebellar neuron network. Proc. of the I.E.E.E. **56**, 960–968 (1968). [37]
— Architecture of the cerebral cortex. In: Basic mechanisms of the epilepsies, ed. by H.H. Jasper, A.A. Ward and A. Pope. Boston: Little, Brown & Co. 1969. [15, 16, 37]
SZILARD, L.: On memory and recall. Proc. nat. Acad. Sci. (Wash.) **51**, 1092–1099 (1964). [26]
TAUB, E.: Prism compensation as a learning phenomenon: a phylogenetic perspective. In: The neuropsychology of spatially oriented behavior. Homewood, Ill.: Dorsey Press 1968. [50]
TEILHARD DE CHARDIN, P.: The phenomenon of man. New York: Harper 1959. [87, 91 92, 96, 152, 162]
TEUBER, H.-L.: Alterations of perception after brain injury. In: Brain and conscious experience, p. 182–216, ed. by J.C. Eccles. Berlin-Heidelberg-New York: Springer 1966. [50]
THOMPSON, H.B.: The total number of functional cells in the cerebral cortex of man, and the percentage of the total volume of the cortex composed of nerve cell bodies, together

with a comparison of the number of giant cells with the number of pyramidal fibres. J. comp. Neurol. **9**, 113–140 (1899). [9, 123]
THORPE, W. H.: Biology, psychology and belief. London: Cambridge University Press 1961. [84]
— Biology and the nature of Man. Riddell Memorial Lectures. Thirty-third Series. London: Oxford University Press 1962. [85, 152]
TÖNNIES, J. F.: Die Erregungssteuerung im Zentralnervensystem. Arch. Psychiat. Nervenkr. **182**, 478–535 (1949). [28]
TURING, A. M.: Computing machinery and intelligence. Mind **59**, 433–460 (1950). [171]
VALVERDE, F.: Apical dendritic spines of the visual cortex and light deprivation in the mouse. Exp. Brain Res. **3**, 337–352 (1967). [33, 36, 37]
— Structural changes in the area striata of the mouse after enucleation. Exp. Brain Res. **5**, 274–292 (1968). [33–35, 37]
WASHBURN, S. L.: The evolution of human behavior. In: The uniqueness of man, ed. by J. D. Roslansky. Amsterdam, London: North-Holland Publishing Co. 1969. [136]
WIESEL, T. N., HUBEL, D. H.: Effects of visual deprivation on morphology and physiology of cells in the cat's lateral geniculate body. J. Neurophysiol. **26**, 978–993 (1963a). [34]
— — Single-cell responses in striate cortex of kittens deprived of vision in one eye. J. Neurophysiol. **26**, 1003–1017 (1963b). [34]
WIGNER, E. P.: Two kinds of reality. The Monist **48**, 248–264 (1964). [46, 48, 59, 61, 63, 66, 152, 172, 178]
— Are we machines? Proc. Amer. Philos. Soc. **113**, 95–101 (1969). [61, 100, 152]
WOLFE, A.: The myth of the free scholar. University Review, vol. 2, p. 3–7. New York: State University of New York 1969. [183]
YOUNG, J. Z.: Growth and plasticity in the nervous system. Proc. roy. Soc. B **139**, 18–37 (1951a). [28]
— Doubt and certainty in science. London: Oxford University Press 1951b. [173]
— A model of the brain. Oxford: Clarendon Press 1964. [86]

Subject Index

Universitätsdruckerei H. Stürtz AG Würzburg